Professor Goodwin is Chairman of Psychiatry at the University of Kansas School of Medicine. His other books for Oxford University Press are *Psychiatric diagnosis* (co-authored with Dr Samuel B. Guze), *Is alcoholism hereditary?*, and *Alcoholism: the facts*.

OXFORD MEDICAL PUBLICATIONS

Phobia

THE FACTS

Phobia

THE FACTS

DONALD W. GOODWIN, M.D.
*Professor and Chairman of Psychiatry
The University of Kansas*

OXFORD   NEW YORK   TORONTO
OXFORD UNIVERSITY PRESS
1983

RC
535
.G65
1983

Oxford University Press, Walton Street, Oxford OX2 6DP
London Glasgow New York Toronto
Delhi Bombay Calcutta Madras Karachi
Kuala Lumpur Singapore Hong Kong Tokyo
Nairobi Dar es Salaam Cape Town
Melbourne Auckland
and associates in
Beirut Berlin Ibadan Mexico City Nicosia

OXFORD *is a trade mark of Oxford University Press*

© *Donald W. Goodwin, 1983*

*All rights reserved. No part of this publication may be reproduced,
stored in a retrieval system, or transmitted, in any form or by any means,
electronic, mechanical, photocopying, recording, or otherwise, without
the prior permission of Oxford University Press*

British Library Cataloguing in Publication Data
Goodwin, Donald W.
Phobia.—(Oxford medical publications)
1. Phobias
I. Title   II. Series
616.85'225   RC535
ISBN 0-19-261395-2

Library of Congress Cataloging in Publication Data
Goodwin, Donald W.
Phobia: the facts.
(Oxford medical publications)
Includes bibliographical references and index.
1. Phobias.   I. Title.   II. Series.
RC535.G65   1983   616.85'225   83-4198
ISBN 0-19-261395-2

Typeset by Colset Private Limited, Singapore
Printed in Great Britain by R. Clay &
Co. Ltd, Bungay, Suffolk

To Caitlin, Mary, and Sarah

Preface

This is a book about fears that don't make sense — fears we know are absurd, fears we wish we didn't have, fears we don't know how we got, fears often too ridiculous or embarrassing to tell even our closest friends about: in short, phobias.

How common are phobias? Nobody knows. There are guesses and statistics, but still nobody knows. Over a lifetime, perhaps most of us experience fears which are a bit absurd. Usually they do little harm and eventually go away. But some phobias hang on. Some do immeasurable harm. Some — rarely — wreck a person's life.

The book opens with a discussion of fear. Like love and rage, fear is a product of the nervous and endocrine systems. 'I will show you fear in a handful of dust', T.S. Eliot wrote. Fear has livelier origins in this book.

Some fears are instinctual and widely distributed in the animal kingdom. These are discussed first, followed by comments about fears that are learned and fears that are unlearned. 'Anything that is learned can be unlearned' is one of the happier axioms in psychology, even if not always true.

Then phobias are discussed — their definition, prevalence, and place in the history of medicine. The convention today is to separate phobias into three types: simple phobias, social phobias, and agoraphobia. Each type is considered separately, although, like most conventions, the separation at times may be artificial. Phobias in childhood — including school phobia — receive special attention.

The focus thus far is on phobic disorders. Phobias also occur as symptoms of other illnesses, such as depression.

Preface

Chapter 7 tells how to diagnose three illnesses in which phobias may be a symptom.

The cause of phobias is unknown, but there are fascinating theories, and some may even be partly true. Chapter 8 describes these theories, starting with Little Hans and Little Albert, names familiar to every student of psychology.

Some phobias go away, others don't. Phobias that don't go away by themselves can sometimes be coaxed into leaving, and Chapter 9 tells how this is done. Phobia therapy lends itself to dubious practices and inflated claims, but there are well-trained people who can, by various means, actually help. The evidence that they help is presented for the benefit of chronic doubters. (Chronic doubters are particularly susceptible to phobias, and may constitute the main audience of this book).

The 'Notes and references' section contains sources for much of the information — something designed to keep chronic doubters safely in a library (assuming it is not a small, crowded library, and no more than two floors off the ground).

To paraphrase Dante: All fear abandon, ye who enter here!

Kansas City   DWG
January, 1983

Contents

Acknowledgements	x
1. Fear	1
2. Phobia: an introduction	24
3. Simple phobias	30
4. Social phobias	44
5. Agoraphobia	55
6. Phobias in childhood	64
7. Phobias in other mental disorders	74
8. The cause of phobias	89
9. The treatment of phobias	105
Notes and references	127
Index	146

Acknowledgements

My thanks to Dr Elizabeth Penick and Dr Ronald Martin, and my wife, Sally, for reading the manuscript and making many helpful suggestions. The secretarial talents of Mrs Evelyne Karson were invaluable. In a book called 'The Facts', there should be no mistakes, but if one or two creep in, the author alone is responsible.

ERRATUM

page 81: The first line on this page should read 'His inability to approach brown objects greatly limited his activities.'

1

Fear

> Why do I yield to that suggestion
> Whose horrid image doth unfix my hair
> And make my seated heart knock at my ribs
> Against the use of nature?
> *Macbeth* I, iii, 134

> They that live in fear are never free.
> Robert Burton
> *The anatomy of melancholy*

Four words are used repeatedly in this book and need defining. They are fear, anxiety, panic, and phobia.

Fear comes from the Old English, *faer*, meaning sudden danger. It refers to fright where fright is justified. The danger is concrete, real, knowable. The fear is appropriate, and sometimes useful, if one is to escape harm.

Anxiety comes from the Latin, *anxius*, meaning a tight feeling in the chest. It refers to a fear of uncertain origin. The person does not know why he is afraid — or his fear seems disproportionate to the danger. 'I just feel anxious', he says, often with a tight feeling in the chest.

Panic refers to extreme fear. The term comes from Pan, the Greek rural deity. Pan is sometimes a friendly god, looking over sheep and shepherds, and a music-lover. But Pan also sometimes scares the hell out of people. In short, Pan is a mixed blessing, and panic, too. Panic may head you straight for the exit in case of fire — but sometimes the wrong exit.

A *phobia* is an intense, recurrent, unreasonable fear, more fully defined in Chapter 2.

To understand phobia, one must know something about fear. Fear originates in the nervous system. Some fears are innate; others learned. Fears tend to go away, given certain

Phobia: the facts

conditions. The following is a review of these four facets of fear:

1. Fear and the nervous system

As far as we know, individuals in all societies have the same emotions, although cultural factors may influence their intensity and expression. In 1872 Darwin observed that animals and humans had a variety of facial expressions of emotion. He believed facial expressions of the various emotions were universal and identical in all cultures.

It seems true: individuals in modern Western and Oriental cultures and primitive tribesmen in New Guinea all show common facial expressions for basic feelings. Here facial expression is more reliable than words. Some languages have no single terms for 'anxiety' or 'depression'. In translating Yoruba into English, for example, recourse has to be made to metaphors such as 'the heart is weak', for depression, and 'the heart is not at rest', for anxiety. One cannot be sure that these phrases correspond at all precisely with the English words.

Not only are some languages richer than others in the language of emotion; even within cultures members of under-privileged groups may have less well-developed abilities to recognize or to express nuances of feeling.

When it comes to fear, the 'language of the body' is the most eloquent and universal. In *The anatomy of melancholy*, (1621), Robert Burton described this well. Fear causes 'many lamentable effects' in men, 'as to be pale, tremble, sweat; it makes sudden cold and heat to come over all the body, palpitations of the heart, fainting . . .' Burton knew fear but nothing about the nervous system. In 1621 nobody knew much about the nervous system.

Now we know that fear begins in the brain, first in the outermost layers of the brain called the cortex. When you hear the word 'Fire!' in a theatre, it is registered in the cortex. But then the 'old brain' takes over — the reptilian brain submerged

Fear

deep beneath the clever new cortex. The old brain does not 'study the situation': it acts, and acts the way it has for millions of years in thousands of species going back at least to the earliest vertebrates. It prepares the body to defend itself.

Here is a good analogy: the President, the cortex, declares a state of emergency; the Defence Department musters the troops; and the troops — billions of cells — pull together for one of two possible actions: fight or flee.

Whichever happens, the messages from the old brain to the body are identical. They travel over two routes. One involves nerve fibres, the other hormones.

There are really two nervous systems. The *skeletal* nervous system activates muscles attached to bones (the skeleton) and makes it possible to run or fight — or just shake, as fear affects some people. The *autonomic* nervous system gives orders to internal organs like the heart and gut and instructs the adrenal glands — the little mushroom-like structures perched on top the kidneys — to release adrenaline, a chemical that resembles the amphetamines. Autonomic means automatic. Once the cortex says 'trouble!' the nervous system starts firing off muscles and glands. Although you may be able to reason with your skeletal muscles — appear calm, stop trembling — the old, primitive autonomic nervous system goes its own way.

The autonomic nervous system has two parts, the *sympathetic* and the *parasympathetic* nervous systems, which have more or less opposite actions. The sympathetic nervous system is the emergency system; the parasympathetic system conserves energy for future emergencies. The first prevails in tornadoes; the latter is business-as-usual.

Picture yourself with a tornado bearing down on you and you can easily predict whether your sympathetic or parasympathetic system is in command:

1. Your pupils dilate: the better to see with.
2. Your heart pumps faster: more blood means more energy.

Phobia: the facts

3. You breathe faster: you need the oxygen.
4. You become pale: the blood is shunted from the skin to the muscles (the better to run with) and to the brain (the better to think with).
5. You stop digesting food: the blood is needed elsewhere, and if there is no tomorrow — the tornado is getting closer — who needs to digest food?
6. You sweat: the sweat evaporates and cools off hot muscles.
7. Your hair stands on end. This does not help your survival chances at all, and merely shows how primitive and out-of-date our autonomic (automatic) responses sometimes are. When a cat's hair stands on end, the cat looks bigger and may scare away enemies, but not even the hairiest of us have that ability. 'It is certainly a remarkable fact', Darwin wrote, 'that the minute muscles, by which the hairs thinly scattered over man's almost naked body are erected, should have been preserved to the present day; and that they should still contract under the same emotions, namely, terror and rage, which caused the hairs to stand on end in those lower members of the Order to which man belongs.'

All these things happen because the sympathetic nervous system takes over in emergencies. When the emergency is over, the parasympathetic system returns to power: your pupils get smaller (the better to read with to learn about future emergencies) and your supper resumes its journey down your 20 feet of small intestine. Also, you start noticing the opposite sex again; sex is one of the first casualties of fear.

Sometimes the fear response outlasts the danger for a time, and the reason is adrenaline. The sympathetic fibres release this amphetamine-like hormone from our adrenal glands and, circulating through the body, it does pretty much what the sympathetic fibres do: keep us primed for danger. The fibres can be switched off immediately; it takes longer for adrenaline to clear from the blood.

Fear

This is how the body handles tornadoes, but what about the *feeling* of fear? The feeling, of course, occurs in the brain. Not only that, it occurs in specific areas in the brain. Put an electrode deep into one of several 'centres' in the brain (as is sometimes done to treat epilepsy), and fear results. Inserted in other centres, and one can produce feelings of anger, depression, and pleasure. You can also inject certain chemicals into these brain centres and produce specific feelings. When epileptics have seizures, they sometimes experience intense fear (or pleasure, or depression), presumably because the seizure originates in a particular feeling centre.

Some cells in the old brain, where fear can be produced with electrical impulses, have receptors with a specific affinity for fear-reducing drugs such as Librium and Valium. The receptors are like locks and the drugs like keys that fit *only* these locks.

This was just learned recently and it raises a very interesting question. Before Librium was synthesized by a Polish chemist in 1955, there were no known chemicals in nature that fit these particular locks. What were they doing there? Was it possible that the brain itself produced drugs like Librium which would be released in emergencies and for which the receptors were designed? Do we have built-in Librium? Do some of us have *more* built-in Librium than others, explaining why some of us are braver than others.

A naturally occurring fear-reducing chemical has not yet been found in the brain, but there is hope that it will be. Pain-reducing drugs, like morphine, also have specific receptors in the brain. Recently, naturally occurring morphine-like chemicals have been found in the brain, explaining why there were morphine receptors long before anyone ever used morphine. It also may explain why some people are more tolerant of pain than others: they have more natural 'morphine'.

All this suggests that bravery may not entirely be due to a

Phobia: the facts

strong character and that the athletes we admire so much on television may not only be blessed with bulging muscles but also with a generous supply of built-in Librium and morphine.

The connection between brain receptors, the feeling of fear, and the physiological changes accompanying fear is beyond comprehension at present. There is a controversy going back a century over which comes first: the feeling of fear or the physiological changes. William James, the Harvard philosopher-psychologist, believed the changes came first. The sense organs and brain perceive a danger. Palpitations, rapid breathing, a tensing of muscles follow. Only then does the person experience fear, a consequence of his recognition of the physical changes signifying danger.

This sounds rather stuffy and academic — who cares which comes first? — except for one thing. If the physiological changes could be eliminated (in situations where they are not needed), then the sensation of fear might be reduced or abolished.

Sure enough, to some extent this can be accomplished, affording some support for the James theory (usually called the James-Lange theory). Drugs that block palpitations have been given to anxious patients, and have made them feel less anxious. Lactic acid, a chemical that produces palpitations, also produces fear, especially in fear-prone people. (Adrenaline also produces palpitations but not often fear.)

Perhaps the strongest evidence for the James-Lange theory is the following: people paralysed from the waist down experience fear and anger but those paralysed from the neck down experience these emotions to a much lesser degree. The same applies to sexual excitement and grief. Quadraplegics describe themselves as acting emotionally but not feeling emotional. According to one, 'I say I am afraid, like when I am going into a real stiff examination at school, but I don't really feel afraid, not all tense and shaky with that hollow feeling in my stomach, like I used to.'

Fear

This raises hope for pathologically fearful people (and the world is full of them) that drugs may be developed specifically to eliminate the physiological fear responses.

Eliminating all fear — through modern chemistry or otherwise — would of course be disastrous. Like such drives as hunger, thirst, and sex, emotions have survival value. Fear is a normal reaction to danger and enhances the survival of both the individual and the species. Psychologists have devised a U-shaped model demonstrating the relationship between fear and performance. In the absence of fear, performance is poor; with some degree of fear, performance improves; with excessive fear, performance once again is poor. By modulating anxiety, we enhance the likelihood of survival.

2. Innate fears

There is no question that animals inherit fears. Consider the famous hawk experiment:

If you pass a V-shaped shadow sideways or backwards over a baby chick in a lab, nothing happens. But if you move the same shadow forward, the baby chick goes into a panic. He has inherited the genetic message — *hawk, danger* — for a certain-shaped shadow, even though he is only a couple of days old and has not talked to his mother and may never see a hawk. According to Konrad Lorenz, the 'father of ethology', the V-shaped shadow is a releasing mechanism for an innate fear response. Let us consider some other innate fear responses.

Fear of darkness

Almost all children are afraid of the dark after the age of three. Apparently what the child really fears are the horrendous creatures — serpents, monsters — conjured up by the imagination. Children in different cultures imagine the same monsters; it is hard to attribute this to upbringing.

Phobia: the facts

Fear of strangers

This occurs normally in infants 6-12 months old. Fear of strangers and smiling are related. By two months the baby smiles indiscriminately. After six months, the baby smiles but only at familiar faces, particularly familiar combinations of forehead, eyes, and nose (incidentally, sex or race do not matter). Seeing both eyes appears to be necessary for the smile response. Strangers elicit screams rather than smiles, particularly if they have large eyes (perhaps wearing spectacles) and large teeth.

Smiling and fear of strangers are universal in 6-12-month old babies. Smiling certainly makes sense. It elicits parental care. It is the first evidence of social interaction in humans, and has survival value. Other mammals and birds also fear strangers. The chimpanzee, for example, begins fearing strangers at about the same time in life as the human infant.

Innate fears are modified by early experience. If the infant has been exposed to many people, he will not be as shy as he would be if he had grown up with only a few (just the family, say). In the latter case, a stranger may produce a response as violent as pain does.

The response is at full strength on first exposure, showing that it is not learned. However, there still is an experiential factor. If the chimpanzee has been reared in darkness, or if the human child has had congenital cataracts until after 6 or 8 months and has then been operated on, the fear response does *not* occur on first exposure. The infant must first become visually familiar with a small group of persons; only afterwards, does a stranger evoke fear.

From these observations the psychologist, Donald Hebb, concluded that fear of the strange is produced by events that combine familiar and unfamiliar. A conflict is involved, he says. He gives the following example:

'Dr R. and Mr T. are regular attendants in a chimpanzee nursery; the infants are attached to both, and evenly welcome being picked up by

Fear

either. Now, in full sight of the infants, Dr R. puts on Mr T.'s coat. At once he evokes fear reactions identical with those made to a stranger, and just as strong. *An unfamiliar combination of familiar things, by itself, can therefore produce a violent emotional reaction.*

In situations where previous learning is an element in the fear response, years may pass before it can occur. Fear of the dark, for example, does not occur until age three or later, allowing time for the brain to be capable of imagining monsters.

Fear of dead or mutilated bodies

This fear can be reduced on repeated exposure; consider the soldier, the medical student, the undertaker's assistant. But as a first visit to the dissecting room shows, the reaction is often powerful on the first exposure, indicating an innate factor. It appears also in milder form with exposure to persons with mutilating injuries (especially to the face).

Again, the reason may be that dead and mutilated people look almost real, but not quite; just as that scary monster Frankenstein looks almost human, but not quite.

Chimpanzees feel comfortable when they are with other chimpanzees but shown the sculptured head or death mask of a chimpanzee they become panic-stricken. The response is age-dependent. Two-year-old chimps ignore the model head, looking only at the experimenter; five-year-olds are fascinated, coming close to stare persistently and excitedly at it; older animals, nine and over, show outright terror, screaming with hair on end. Even with a wire cage between them and the reasonably faithful clay model of a chimpanzee head, none of the older animals will approach it.

Fear of snakes

Adult chimpanzees fear snakes. They fear snakes even if they have never seen a snake before. The fear is inherited.

Do humans have an inherited fear of snakes? Maybe. If so, it comes with age: small children do not usually fear snakes. It

Phobia: the facts

may not develop at all if the child has frequent opportunities to play with snakes; they learn *not* to fear snakes. But if you show a snake to city-bred adolescents who have never seen a snake and tell them it is totally harmless, they may say they believe you, but they still keep their distance.

Perhaps the city-bred kids have *learned* that snakes are dangerous from reading or from their parents. Still, adolescents learn about a lot of dangerous situations — such as taking drugs and driving too fast — and this does not seem to discourage other forms of risk-taking.

Also, it has been suggested that perhaps chimpanzees and city children exposed to a snake for the first time are not frightened by the snake itself (they do not have a small portrait of a snake labelled 'danger' in their brains) but rather are fearful of the writhing movements of the snake. There is, in fact, evidence that chimps fear strange *moving* objects more than similar stationary objects.

Still, fears of strangers, serpents, and snakes are pervasive and most authorities suspect a hereditary factor. One early authority, Stanley Hall, in 1897, noted that

serpents are no longer among our most fatal foes. Most of the animal fears do not fit the present conditions of civilized life; strangers are not usually dangerous nor are big eyes and teeth . . . yet, the intensity of many fears, especially in youth, is out of all proportion to the exciting cause. First experiences with water, the moderate noise of the wind, distant thunder, etc., might excite faint fear, but why does it sometimes make children frantic with panic?

Adam Smith says it is in the genes:

There was a time when humanity slept by the fire and the predators really were there in the dark. One cell carries the instructions on how to make a whole new person: let's see, make the eyes brown, make the ears thus, make the nose thus, let's set the trigger for adolescent growth at 12.2 years, and oh, yes, let's throw in the message about the bears, predators by the fire, for ages five to nine.

Fear of dark woods

When people started using LSD in the early 1960s, they shared

Fear

their experiences and learned they were often the *same* experiences. One common LSD-produced fear was of entering dark woods and not finding the way back. As one user said, 'Everybody had a dark woods in the middle of their head, and sooner or later they got around to it. So when Bob Dylan sang about the foggy ruins of time, past the frozen leaves, the haunted, frightened trees, he touched a universal of sorts.'

The feeling goes back at least to the time when dark woods really were dangerous and that is where most people lived, a time resurrected by fairy tales and Walt Disney movies. Whether the fear is innate or learned (or both), dark woods are a pervasive symbol for lurking monsters and the unknown. Dante opens *The Divine Comedy* with 'In the middle of the journey of our life, I found myself in a dark wood', and a terrifying place it turns out to be.

Fear of heights

Even goats fear heights. This was shown in a classical experiment. A board was laid across the centre of a sheet of glass. On one side of the board a sheet of patterned material was placed flush against the undersurface of the glass, giving the glass the appearance of solidity. On the other side a sheet of the same patterned material was laid upon the floor 1 ft below the glass. Placed on the 'deep' side of the glass, newborn goats froze; they relaxed again only when removed to the 'shallow' side.

Human infants behaved the same way, crawling readily on the 'shallow' side of the board and avoiding the 'deep' side. Fear of a receding edge remains among adults on the edge of a precipice, with a feeling of being drawn down and a protective reflex to withdraw from the edge.

Fear of being looked at

Eyes are probably the first thing babies notice. They are small, colourful, move, and reflect light. Two eyes, as noted, are required to elicit the first social response, smiling. The first drawings of children often are of big round heads with eye

Phobia: the facts

spots. Legs and arms are added later. 'Fear of two staring eyes is ubiquitous throughout the animal kingdom, including man', according to Isaac Marks, a leading authority on phobias. Belief in the power of the look seems universal and independent of culture. In man, large staring eyes are used in defensive magic. Many species of mammals and birds use eyes and eye markings as threat displays and defence against attack.

Stare at a monkey in a cage and you produce erratic changes in the monkey's brain waves as well as his behaviour. Stares produce discomfort in most of us, and the cause seems to be inheritance as well as early-life experiences.

Fear of novelty

People and animals have mixed responses to novelty. The novel can produce fear but may also be sought out. Novelty can attract and repel in turn, as demonstrated by Lorenz's wonderful description of a mixed-up raven.

A young raven, confronted with a new object, which may be a camera, an old bottle, a stuffed polecat, or anything else, first reacts with escape responses. He will fly up to an elevated perch and from this point of vantage, stare at the object . . . maintaining all the while a maximum of caution and the expressive attitude of intense fear. He will cover the last distance from the object hopping sideways with half-raised wings, in the utmost readiness to flee. At last, he will deliver a single fearful blow with his powerful beak at the object and forthwith fly back to his safe perch. If nothing happens he will repeat the same procedure in much quicker sequence and with more confidence. If the object is an animal that flees, the raven loses all fear in the fraction of a second and will start in pursuit instantly. If it is an animal that charges, he will either try to get behind it or, if the charge is sufficiently unimpressive, lose interest in a very short time. With an inanimate object, the raven will proceed to apply a number of further instinctive movements. He will grab it with one foot, peck at it, try to tear off pieces, insert his bill into any existing cleft and then pry apart his mandibles with considerable force. Finally, if the object is not too big the raven will carry it away, push it into a convenient hole and cover it with some inconspicuous material.

Fear

Monkeys do likewise. Confronted with a strange object, a monkey freezes and stares at the object from a distance. After a while the animal tentatively approaches the object, looking at it, sniffing, touching, and finally handling it. After an hour or so, the animal returns to staring fixedly at the object.

Do our children do the same? In their own way, yes.

Individuals and species vary enormously in their response to fear-provoking situations, even when the response appears largely innate. This variability can be traced to a combination of four factors:

(1) The evolutionary scale

Professor Donald Hebb writes:

As we go from rat to chimpanzee (from lower to higher animals), we find an increasing variety in the causes of fear. Pain, sudden loud noise and sudden loss of support are likely to cause fear in any mammal. For the laboratory rat we need add only strange surroundings, in order to have a list of things that disturb the animal under ordinary circumstances. With the dog, the list becomes longer: we must add strange persons, certain strange objects (a balloon being blown up, for example, or a large statue of an animal) or strange events (the owner in different clothing, a hat being moved across the floor by a thread which the dog does not see). Not every dog is equally affected, of course, but dogs as a species are affected by a much wider variety of things than rats.

Monkeys and apes are affected by a still wider variety, and the degree of disturbance is greater. Causes of fear in the captive chimpanzee make up an almost endless list: a carrot of an unusual shape, a biscuit with a worm in it, a rope of a particular size and color, a doll or a toy animal . . .

And man? The evidence suggests that the more intelligent the animal, the greater the susceptibility to baseless fears. If the next eight chapters of this book are any indication, man must surely surpass any other species in range and variety of baseless fears.

Phobia: the facts

(2) Genetic differences

Species differ greatly in fearlessness: lions and tigers are more fearless than deers and rabbits. Moreover, animals can be inbred to show greater or lesser fear responses to specific stimuli and to be more or less fearless in general. In humans, athletic ability runs in families and is almost certainly influenced by heredity. Further evidence: identical twins, sharing the same genes, fear strangers about equally; fraternal twins, with different genes, have differing degrees of fear.

(3) Imprinting

Experiences in infancy modify innate fear responses. This is called imprinting. Imprinting can occur very early in life. For instance, chicks hatched from eggs incubated in darkness differ from chicks hatched from eggs incubated in light: they are less fearful! From the moment of conception, genetic and environmental influences interact. The question, 'What is inherited? What is learned?' is often unanswerable.

(4) Stimulus characteristics

In animal research, fear is inferred from behaviour. If an animal approaches an object, it is 'unafraid'; if it avoids an object it is 'afraid'. Approach-avoidance studies have produced the following generalizations:

(a) Animals fear (avoid) loud, irregular noises and large objects moving toward them at high speed (particularly objects with sharp, angular corners).

(b) Animals are unafraid of (approach) small, rounded objects which move slowly and make soft, regular noises.

In human terms, the boss who produces fear in his employees is the burly fellow with craggy features and a loud staccato voice who comes at you full-steam, lays both hands on your shoulders, and says . . . it hardly matters what he says. It could be 'Nice day!' and the pupils of the employee will still dilate, his heart pound, and the adrenaline flow.

Fear

(5) Summation

Some situations are feared more than others, not because of any single factor described above but from a combination of factors which, in a given individual in a particular situation, produces fear.

Genetically-determined fear may be 'latent', expressed only with the addition of imprinting or later stressful life experiences or in the presence of unusually intense stimuli. This is called summation. The fearful person is almost never aware of summation: the complex and mysterious reasons why he is afraid.

3. Learned fears

Learning, as used here, refers to conditioning. It can be argued that all learning can be reduced ultimately to conditioning, but whether this is true or not, conditioning, narrowly defined, has a profound relationship to fear.

Our knowledge of conditioning began with the great Russian physiologist, I.P. Pavlov. He rang a bell just before feeding a dog. He did it repeatedly. After a time the dog salivated whenever the bell rang, whether he received food or not. The dog was *conditioned* to salivate.

Four terms are needed to understand conditioning:

Unconditioned stimulus: hunger
Unconditioned response: salivation
Conditioned stimulus: the bell
Conditioned response: salivation

Before learning (conditioning) can occur, there must be an organism and the organism must have 'drives', such as hunger. Drives evolved in the interest of survival of the individual and the species. Drives have one aim: their own abolition. Saliva digests food, temporarily abolishing hunger.

By repeatedly pairing a neutral stimulus (bell) with a drive (hunger), a conditioned response (salivation) occurs which

Phobia: the facts

becomes independent of the original drive. The bell rings, the dog salivates, whether it is hungry or not. The dog has 'learned' something, in much the same way we learn to drive a car, mainly by acquiring a set of conditioned responses.

We constantly form new conditioned responses and discard old ones, and are usually never the wiser.

Habits are conditioned responses. Tying our shoelaces, brushing our teeth, waving at a friend: all are habits, all learned, all more or less automatic, all rooted in a drive with survival value (staying warm, being attractive, maintaining allies).

Conditioned stimuli become paired (and unpaired) to drives endlessly; and although there is a limited number of drives (hunger, sex, etc.), literally *anything* can become a conditioned stimulus.

Say it thunders every time you eat strawberries. After a few days, thunder makes you hungry for strawberries. You may even *crave* strawberries. Your strawberry addiction results from a conditioned response (hunger) to a conditioned stimulus (thunder) which became paired through happenstance. Thus, when one expert says that addictions are biological and another says they are learned, both are right (and neither usually mentions happenstance).

In the same sense, few fears are purely innate or purely learned — at least in higher animals. Four centuries ago, Sir Francis Bacon wrote: 'Men fear Death, as children fear the dark; and as that natural fear in children is increased with tales, so is the other.' Tales become conditioned stimuli, but a biological factor — the innate fear, the unconditioned response — must be present for conditioning to occur.

Words become conditioned stimuli. Think of 'I love you' or 'You're fired', or 'I'll kill you' and the point is obvious.

Non-verbal symbols become conditioned stimuli. A swastika once inspired alarm in entire communities. Secondary sexual characteristics in the human female — breasts, bottoms — invoke unconditioned responses in the human

Fear

male, but the size of breasts and bottoms which invoke these responses changes from era to era, showing how condition-bound unconditioned responses become.

When the unconditioned response is especially strong, a single pairing with a neutral stimulus may produce a lasting conditioned response. A survivor of a plane crash may panic at the mention of the word 'plane'. A mouse may once become ill from poisoned food in a cupboard and never go back to that cupboard again. This is called one-trial learning.

Conditioned stimuli tend to generalize: things remind people of things. Mark Keller, writing about alcoholism (where conditioning plays an important role), describes the phenomenon aptly:

> For the alcoholic, there may be several or a whole battery of critical cues or signals. By the rule of generalization, any critical cue can spread like the tentacles of a vine over a whole range of analogs, and this may account for the growing frequency of bouts, or for the development of a pattern of continuous inebriation. An exaggerated example is the man who goes out and gets drunk every time his mother-in-law gives him a certain wall-eyed look. After a while he has to get drunk whenever any woman gives him that look.

A little boy is frightened by a mouse, later fears any small furry animal, and as a teenager develops a phobia toward all animals. The 'rule of generalization' is important in the development of certain phobias, as later chapters will show.

Confusing stimuli can also produce 'neuroses' like phobias. Pavlov created the first neurotic dog (at least in a laboratory). He did it as follows:

A circle was projected on the wall in front of the dog and then the dog was given food. When an oval was projected on the wall he was not given food. At first the circle and oval were quite different, the ratio of length to breadth in the oval being 2:1. The dog secreted saliva to the circle, not to the oval. Then the ratio was decreased. When it became 9:8 (the oval was now nearly a circle), the dog had a nervous breakdown. He howled, defecated, and refused to enter the experimental room.

Phobia: the facts

Even when the experimenter restored the symbols to their original shapes — perfect circles, unmistakable ovals — the dog remained wacky. 'His neurosis seems permanent!' marvelled Pavlov.

Remember: the dog had not been punished; he was never in pain. He simply was a victim of conflicting signals — one of the more awful things that happen to dogs and to people.

So far, we have been talking about Pavlovian or 'classical' conditioning: the pairing of a neutral stimulus with an unconditioned response. The response — salivation, rapid heartbeat, fear, sexual arousal — is largely involuntary.

There is another type of conditioning called 'operant' conditioning where the animal or person learns to behave in ways calculated to produce pleasure or avoid pain. The behaviour is often repetitive and predictable — this is why it is called 'conditioning' — but the person has more control over his behaviour than he does of his salivary glands, and thus operant conditioning is more voluntary and thought-out than classical conditioning. The person is more his own operator, so to speak.

Classical conditioning and operant conditioning are closely related. The first often precedes the second. For example:

Put a rat on an electrified grid. Flash a red light, and one-half second later, give him an electric shock. After doing this for a time, the red light alone will produce freezing, jumping, or other conditioned fear responses. Classical conditioning has occurred.

Then give the rat an opportunity to avoid or escape the shock. Eventually the fear response to red lights will disappear. If, for example, he learns that by pressing a bar when the light flashes, the shock will not occur, after a while the rat becomes a skilled bar-presser. He has learned to avoid punishment by his behaviour. Should he become lazy or forget to press the bar, another jolt serves as a reminder. The jolt is called a 'negative reinforcer'.

How many negative reinforcers in the form of memoranda flow daily from bosses to employees?

Fear

It also works the other way. If an animal learns to press a bar to receive food, the food, a reward, serves as a positive reinforcer, and the rat becomes a happy and hardworking bar-presser. A nice word from a supervisor often has the same effect: the recipient is not only grateful but works harder.

Behaviour is powerfully shaped by negative and positive reinforcers. We simply learn which actions bring rewards and which bring punishments and behave accordingly. However, it is not always that simple.

Sometimes the *same* action produces both rewards and punishments. It happens all the time in real life and is easy to show in the laboratory.

A rat is first taught to press a bar to avoid shock and later taught to press the same bar to obtain food. You now have a very confused rat: another case of conflicting signals, now compounded by conflicting reinforcers. Clearly a set-up for 'emotional problems' — in rats and people. You love your wife but she is a bad housekeeper: stay or leave? (or both — take up golf).

Conflict is the rule of life. This is partly the basis of the popularity of sedatives and alcohol. Both reduce conflict — temporarily. Again, you can show it in the lab.

Cat-food is placed at the end of a tunnel. A green light flashes when the food is available. At the end of the tunnel is also an electrified grid. A red light flashes when the grid is activated. Both lights become conditioned stimuli: the cat salivates at green and is terrified by red.

Place a very hungry cat in a tunnel and flash both lights simultaneously. Something called an approach–avoidance conflict results. The cat moves a certain distance down the tunnel toward the food but freezes at the expectation of shock. Hunger versus fear. How far he goes down the tunnel will depend on which is stronger: hunger or fear.

At this point give the cat alcohol or a barbiturate. The fear is reduced but not the hunger. He now approaches more than avoids. Reducing the fear pharmacologically has temporarily resolved the conflict.

Phobia: the facts

It happens all the time in bars. The man on the bar stool, the woman sitting at the table . . . the third drink, the fourth . . . the approach. 'Love casts out fear', as Aldous Huxley said. Love and alcohol combined *really* cast out fear.

Animals quickly learn the fear-reducing effects of alcohol and barbiturates. In approach-avoidance situations, they will self-administer both drugs. People learn, also. There is approximately one bar for every grocery store in most towns.

Alcohol reduces fear but the fear does not stay reduced after the alcohol is gone. There are other ways to reduce fear where the fear does stay reduced. They will be discussed briefly now and much more fully in the chapter on the treatment of phobia.

4. Fears that go away

There is an old saying in psychology that *anything learned can be unlearned*. In this message lies hope for fearful mankind — if only more was known about how to unlearn.

Some things are known.

Conditioned stimuli lose their strength (extinguish) if not maintained by at least an occasional reward or punishment.

Pavlov's dog eventually stops salivating on hearing a bell if presentation of food *never* follows the bell. Fear of snakes (even if partly innate) is overcome with frequent exposure to snakes. A flashing neon bar sign remains a conditioned stimulus signalling 'thirst' only as long as it regularly produces the conditioned response of entering the bar and ordering a drink. If the response is suppressed for a sufficient period of time, the stimulus loses its grip.

Extinguishing thirst-producing stimuli, by the way, is an essential part of the treatment of alcoholism, and the difficulty in treating alcoholism is that 'sufficient' time may be a long time indeed.

Suppression of the conditioned response must be unrelenting to neutralize the conditioned stimulus. Erratic responding,

Fear

if anything, strengthens the 'habit'. Slot-machines — and all forms of gambling, for that matter — are popular and even become addictive because the payoff is erratic. 'Maybe this time I'll win' keeps the gambler gambling. It is a thought that never occurs in situations when you never win. Nor does it occur in situations where you *always* win. In the latter, people sometimes stop what they are doing out of sheer boredom.

Conditioned fear stimuli resist extinction under two circumstances. First, if the person avoids exposure to the feared situation, he never has an opportunity to unlearn the fear. Second, if alcohol or drugs are used to reduce the fear, they do so, as mentioned, only temporarily. As all drinkers know, the fear returns — often redoubled — on the morning after.

Non-responding to fearful stimuli and familiarization with fearful situations are ways people handle ordinary fears all the time. Familiarization often is vicarious. Hearing someone describe a fear they have had and how it was overcome can be reassuring. Just knowing that others have fears helps. Some people — a stoical woman, the macho man — may *seem* fearless, making others feel weak and cowardly, but it is often camouflage. Movies, plays, novels: all depict fear in others and help us handle ours.

Familiarization with fear through shared experiences often takes the form of gossip. Gossip is considered a minor vice, whereas, on the contrary, it is one of the nicest things we do. The novelist Dan Davin explains why:

Gossip, talking behind people's backs, is a necessary ingredient of our social culture, indispensable to it and at its best one of the finest instruments of our civilized living. We all know that we are talked about but . . . we are able to be unaware of it most of the time as we are, happily, unaware of death. We can gossip with enjoyment about others, serenely forgetful that they at the very same moment may be gossiping with equal enjoyment about us.

For gossip is a necessary exercise: it is the practice of our skills in social perception, the daily do-it-yourself art by which we refine and sharpen our sensibilities, subtilize our knowledge of others, check our knowledge by the competing perceptions of our cronies, remind

Phobia: the facts

ourselves of the relativity of all human truth and the exhilarating improbabilities of behaviour.

Through gossip we acquire the sense of others without which we fail as social beings and which is the *sine qua non* of charity.

So fears that are learned can be unlearned, but sometimes you have to work at it and some are better at doing it than others.

There is another principle in psychology that helps people deal with fear. This is the 'law of reciprocal inhibition'.

Reciprocal inhibition simply means that you cannot experience two opposing emotions simultaneously. You cannot be afraid and feel sexy at the same time. (Think so? Try it.) You cannot feel affectionate toward your lover and be mad at him or her simultaneously. Angry men forget their fear in combat. Even Pavlov's neurotic dog showed less fear when sexually stimulated.

Knowingly, unknowingly, people use this principle to combat fear all the time. The fear of flying is overcome by dwelling on the good time awaiting at the other end. Claustrophobes, heart pounding, take the elevator up to the office thinking fervently of the things they will buy with their pay. Fear and socializing do not go together, perhaps explaining why some people are so gregarious.

People learn what comforts them, and learn that what comforts them reduces fear. A little boy puts his thumb in his mouth every time he's frightened. As long as he has his thumb in his mouth he seems impervious to fear. Another child fears certain objects while sitting on the floor but not when sitting in his familiar high-chair.

The thumb and high-chair for these children are a *soteria*, a term for any object or situation from which people derive disproportionate comfort. A soteria is the opposite of a phobia. Examples are stuffed animals and toys which children carry around with them, and charms which many adults wear. For the youngster on the streets, the loud stereo he is lugging

Fear

around may in fact be as much a soteria as it is a radio.

Some people afraid of fainting carry a bottle of smelling salts, although perhaps they have never fainted in their life. Others have aspirin or Valium with them at all times and never take them.

Eating reduces fear. Chimpanzees lose their fear of snakes if bananas are placed on the lid of a glass box containing a snake. Joseph Wolpe, who devised a successful treatment for phobias, discovered the treatment from a study involving food and cats. He put the cats in a cage containing food. Whenever they walked toward the food, they were shocked. Pretty soon the cats decided they would rather go hungry than be shocked, and developed a phobia, so to speak, about food in cages. The phobia persisted even when the shocks were discontinued.

Wolpe discovered a way to 'cure' the phobia which he later applied to human phobias. He first placed the food some distance away from the cage, sufficiently far for the approach tendency due to hunger to overcome the avoidance tendency due to fear. Gradually the food was placed near the cage and finally inside it, until eventually the cats entered the cage and ate the food. More about Wolpe and his phobia cure in Chapter 9.

To some extent, motion and e-motion are incompatible. This has been known for a long time, although the joggers who dot the landscape seem to believe they have discovered something new.

Some people conquer their fear by doing precisely what they fear doing — or something like it. One person fears heights and becomes an acrobat. Another fears water and becomes a sailor. A third fears blood and becomes a surgeon. They develop, in short, *counterphobias*, often hailed as courage but courage with uncourageous roots.

2

Phobia: an introduction

> Love casts out fear; but conversely fear casts out love.
> And not only love. Fear also casts out intelligence, casts
> out goodness, casts out all thought of beauty and truth
> ... in the end fear casts out even a man's humanity.
> Aldous Huxley in *Ape and essence*

A phobia is a persistent, excessive, unreasonable fear of a specific object, activity, or situation that results in a compelling desire to avoid the dreaded object, activity, or situation. The fear is recognized by the individual as excessive and unreasonable. The avoidance behaviour involves some degree of disability.

The definition has four essential features. Phobias are *persistent*. Fear of remounting a horse after falling off one is not a phobia unless the fear persists and leads to the second essential feature: *avoidance* of the feared object, activity, or situation (in this case, horses). Fears which do not lead to avoidance usually go away, assuming they are unrealistic. This fortunate fact of life, indeed, is the basis for most treatments of phobia.

The third element: to be phobic, the fear must not only seem *unreasonable* to others but be viewed as unreasonable by the victim. When a schizophrenic has an irrational fear, he usually does not recognize its irrationality, in which case the fear is not a phobia but a delusion (defined by psychiatrists as a fixed, false idea).

The fourth element is decisive in separating 'normal' fears from phobias. How *disabling* is the fear? Irrational avoidance of objects, activities, or situations is common. If the effect on life adjustment is insignificant, it little matters. Many people fear and avoid harmless insects and spiders, but it does not

Phobia: an introduction

affect their lives. However, when fear and avoidance significantly interfere with a person's mental functioning or social adjustment, it is excessive and therefore phobic.

The four elements of a phobia are illustrated in this account by a phobic psychiatrist:

I was pampering my neurosis by taking the train to a meeting in Philadelphia. It was a nasty day out, the fog so thick you could see only a few feet ahead of your face, and the train, which had been late in leaving New York, was making up time by hurtling at a great rate across the flat land of New Jersey. As I sat there comfortably enjoying the ride, I happened to glance at the headlines of a late edition which one of the passengers who had boarded in New York was reading. 'TRAINS CRASH IN FOG', ran the banner headlines, '10 DEAD, MANY INJURED'. I reflected on our speed, the dense fog outside, and had a mild, transitory moment of concern that the fog might claim us victim, too, and then relaxed as I picked up the novel I had been reading. Some minutes later the thought suddenly entered my mind that had I not 'chickened out' about flying, I might at that moment be overhead in a plane. At the mere image of sitting up there strapped in by a seat belt, my hands began to sweat, my heart to beat perceptibly faster, and I felt a kind of nervous uneasiness in my gut. The sensation lasted until I forced myself back to my book and forgot about the imagery.

I must say I found this experience a vivid lesson in the nature of phobias. Here I had reacted with hardly a flicker of concern to an admittedly small but real danger of accident, as evidenced by the fog-caused train crash an hour or two earlier. At the same time I had responded to a purely imaginary situation with an unpleasant start of nervousness, experienced both as physical symptoms, and as an inner sense of indescribable dread so characteristic of anxiety. The unreasonableness of the latter was highlighted for me by its contrast with the absence of concern about the speeding train, which if I had worried about it, would have been an apprehension grounded on real, external circumstances.

Because a phobia involves a cluster of elements, it is called in medical parlance a 'disorder'. Actually, there are three major phobic disorders: *simple phobia, social phobia,* and *agoraphobia*. Each has the essential four features of a phobia and also what doctors call a characteristic 'natural history'.

Phobia: the facts

This means the disorder has a specific age of onset, course, and outcome. In this sense, a phobic disorder is like measles: there is information about when it begins, what it does, and how it ends. Is a disorder the same as a disease? If one defines disease as a distressing condition which leads people to consult physicians, it meets the definition of disease. People have trouble viewing phobia as a disease because there is no rash or fever and it all seems to be in the person's mind. There is no point in arguing over words. Phobia has at least achieved the status of disorder, which is a step forward from the days when witches were blamed.

Prevalence

How common is phobia? There are two ways to find out. One is to study a random sample of the population. The other is to study people who go to doctors. The latter method is unsatisfactory. People see doctors for other reasons than their illness. They see doctors because doctors are available, because they have the money to see doctors, because seeing doctors is fashionable, or for other reasons. Nevertheless, treatment statistics are the main basis for judging how many people have a particular illness in the general population, however misleading this may be.

There has been only one study of phobia in a random sample of the population. A household survey in Vermont, USA, in 1969 found a prevalence of all phobias of 8 per cent, with severely disabling phobias having a prevalence of 0.2 per cent. 'Severely disabling' was defined as absence from work for an employed person and inability to manage common household tasks for a housewife. It would be useful to distinguish phobias as a *symptom* from phobic *disorders*, but this is rarely done. Less rigorous studies than the Vermont study have estimated the prevalence of phobic *symptoms* of any kind in the general psychiatric population between 20 and 44 per cent.

Phobia: an introduction

Reports of phobias in psychiatric practice are contradictory. One study reported that phobias were rarely seen for psychiatric treatment, representing 2 or 3 per cent of cases in America and England. About half of these cases, mostly women, had agoraphobia. However, the presence of agoraphobia and social phobia in two large series of psychiatric patients with predominant mood disorders has been reported to range from 50 to 65 per cent, and psychiatrists see large numbers of patients with 'predominant mood disorders'. One of these studies had a control group consisting of patients from a fracture clinic where the prevalence of agoraphobia and social phobia was found to be 30 per cent and 16 per cent, respectively. However, these figures may have been spuriously elevated due to the frequency of alcohol abusers in a fracture clinic population. One study found that alcohol abusers have a particularly high rate of phobia, one-third suffering from disabling phobias and another third from milder phobias.

There are two reasons for the discrepancy in prevalence data about phobias: (1) some physicians may include what others call 'normal fears' in their statistics and (2) people tend to be secretive about phobias. Many people are embarrassed about their phobias and often keep them secret from even their closest friends. This habit of secrecy may persist even when they see doctors and see them for psychiatric reasons. It is not unusual in psychiatric practice to see a patient for a long period and then have him describe, almost in passing, a phobia that has plagued him for years. In routine questioning of patients, phobias often are not asked about. Many non-psychiatrists seem unaware even of their existence, or at least of their possible medical significance.

If only 8 per cent of adults have at least a somewhat disabling phobia, this means the problem affects a large number of people. Fortunately, most phobias are 'benign' in that most people eventually recover from them whether treated or not.

Phobia: the facts

Historical background

Phobos was a Greek god called upon to frighten one's enemies. His likeness was painted on masks and shields for this purpose. Phobos, or phobia, came to mean fear or panic.

'Phobia' first appeared in medical terminology in Rome 2000 years ago, when 'hydrophobia' (fear of water) was used to describe a symptom of rabies. Though the term was not used in a psychiatric sense until the nineteenth century, phobic fears and behaviour were described in medical literature long before that. Hippocrates described at least two phobic persons. One was 'beset by terror' whenever he heard a flute, while the other could not go beside 'even the shallowest ditch' and yet could walk in the ditch itself.

Robert Burton in *Anatomy of melancholy* distinguishes 'morbid fears' from 'normal' fears. Demosthenes' stage-fright was normal, while Caesar's fear of sitting in the dark was morbid. Burton believed normal fears could be overcome by will power, but not morbid fears.

The term phobia appeared increasingly in descriptions of morbid fears during the nineteenth century, beginning with 'syphiliphobia', defined in a medical dictionary published in 1848 as 'a morbid dread of syphillis giving rise to fancied symptoms of the disease'. Numerous theories were advanced to explain phobias, including poor upbringing.

In 1871, a German neurologist named Karl Westphal described three men who feared public places and labelled the condition agoraphobia, '*agora*' coming from the Greek word for place of assembly or marketplace. Historians give Westphal credit for first describing phobia in terms of a *disorder* rather than an isolated symptom. Westphal even prescribed a treatment for the condition, suggesting that alcohol, a companion, or the use of a walking cane in public places would be helpful.

Later investigators compiled long lists of phobias, naming each in resounding Greek or Latin terms after the object or

Phobia: an introduction

situation feared. Thus, as the contemporary psychiatrist John Nemiah points out, 'the patient who was spared the pangs of taphaphobia (fear of being buried alive) or ailurophobia (fear of cats) might yet fall prey to belonophobia (fear of needles), siderodromophobia (fear of railways), or triskaidekaphobia (fear of thirteen at table), and pantaphobia was the diagnostic fate of that unfortunate soul who suffered from them all.'

Since the late nineteenth century there has been a continuing controversy over the relationship of phobias to other psychiatric disorders. The famous German psychiatrist, Emil Kraepelin, spoke of phobias and obsessions as though synonymous. The even more famous Sigmund Freud separated phobic neurosis from obsessive-compulsive neurosis and anxiety neurosis — a separation that still prevails in most psychiatric textbooks. Some psychiatrists regard all phobias as a manifestation of manic-depressive disease. The majority, however, agree that phobias may occur in various psychiatric conditions, but may also be the primary manifestation of specific phobic disorders.

This trend is reflected in the inclusion of 'Phobic Disorders' in the current American Psychiatric Association classification of psychiatric disorders. It is also reflected in the organization of the present book, which first considers the three major types of phobic disorders and later discusses phobias as manifestations of other psychiatric disorders.

3

Simple phobias

> There were 117 psychoanalysts on the Pan Am flight to Vienna and I'd been treated by at least six of them . . . God knows it was a tribute either to the shrinks' ineptitude or my own glorious unanalyzability that I was now, if anything, more scared of flying than when I began my analytic adventures some thirteen years earlier.
>
> My husband grabbed my hand therapeutically at the moment of takeoff.
>
> 'Christ — it's like ice', he said . . . My fingers (and toes) turn to ice, my stomach leaps upward into my rib cage, the temperature in the tip of my nose drops to the same level as the temperature in my fingers . . . and for one screaming minute my heart and the engines correspond . . . I happen to be convinced that only my own concentration . . . keeps this bird aloft . . . I congratulate myself on every successful takeoff, but not too enthusiastically because it's also part of my personal religion that the minute you grow overconfident and really *relax* about the flight, the plane crashes instantly . . .
>
> Erica Jong in *Fear of flying*

A simple phobia is simply that: an isolated fear of a single object or situation, leading to avoidance of the object or situation. The fear is irrational and excessive but not always disabling because the object or situation can sometimes be easily avoided (e.g. snakes, if you are a city dweller).

Impairment may be considerable if the phobic object is common and cannot be avoided, such as a fear of elevators by someone who must use elevators at work. Fear of flying — the bane of Erica Jong and the phobic psychiatrist in the preceding chapter — is, to say the least, inconvenient in the jet age, and devastating for national politicians and travelling salesmen.

Simple phobias

Simple phobia is not usually associated with other psychiatric symptoms or other psychiatric disorders, such as depression. The phobic person is no more (or less) anxious than anyone else until exposed to the phobic object or situation. Then he becomes overwhelmingly uncomfortable and fearful, sometimes having symptoms associated with a panic attack (palpitations, sweating, dizziness, difficulty breathing). The phobic person can also fear just thinking about the possibility that he might run into or be confronted with the phobic stimulus. This is called anticipatory anxiety; it leads people to avoid all sorts of situations in which the phobic stimulus *might* be present.

Simple phobia rarely leads to medical consultation. Except for a single small household survey, there have been no epidemiological studies of simple phobias, so nobody knows how common they are. There is one reason to believe simple phobias are at least fairly common. When a new and apparently effective treatment for simple phobias was introduced by Joseph Wolpe in 1958, people with simple phobias started showing up in doctor's offices. When the Maudsley Hospital in London ran newspaper advertisements announcing a phobia clinic, patients came in hordes. Some did not have a phobia or much of a phobia; they were lonely and wanted to talk to someone; or they had a depression or something else wrong and did not know who to see. But many had true phobias they had never mentioned to anyone, much less to a doctor.

Psychiatrists often have the experience of a phobia suddenly being introduced by a patient being treated for something else.

If phobias were as common as *names* for phobias, they would outnumber common colds. The reason for so many names is this: people can become phobic about almost any object or situation. Classifying phobias by the feared object or situation — a common practice in the nineteenth century — can be performed by anyone with a little Greek or Latin, as

Phobia: the facts

shown in the table (compiled by Isaac Marks to give a taste of what is possible).

Formal names which have been given to some phobias

Acrophobia	:	height (Gr. *acro*, heights or summits)
Agora—	:	open spaces (Gr. *agora*, market place, the place of assembly)
Ailuro—	:	cats (Gr. *ailuros*, cat)
Arachno—	:	spiders (Gr. *arachin*, spider)
Antho—	:	flowers (Gr. *anthos*, flower)
Anthropo—	:	people (Gr. *anthropos*, man generically)
Aqua—	:	water (Lat. *aqua*, water)
Astra	:	lightning (Gr. *asterope*, lightning)
Bronto— }	:	thunder (Gr. *bronte*, thunder)
Keraunos— }	:	(Gr. *keraunos*, thunderbolt)
Claustro—	:	closed spaces (Lat. *claustrum*, bar, bolt, or lock)
Cyno—	:	dogs (Gr. *cynas*, dog)
Demento—	:	insanity (Lat. *demens*, mad)
Equino—	:	horses (Lat. *equus*, horse)
Herpeto—	:	lizards, reptiles (Gr. *herpetos*, creeping or crawling things)
Mikro—	:	germs (Gr. *mikros*, small)
Muro—	:	mice (Lat. *murmus* mouse)
Myso—	:	dirt, germs, contamination (Gr. *mysos*, uncleanliness, abomination)
Numero—	:	number (Lat. *numero*, number)
Nycto—	:	darkness (Gr. *nyx*, night)
Ophidio—	:	snakes (Gr. *ophis*, snake)
Pyro—	:	fire (Gr. *pyr*, fire)
Thanato—	:	death (Gr. *thanatos*, death)
Tricho—	:	hair (Gr. *tricho*, hair)
Xeno—	:	stranger (Gr. *xeros*, stranger)
Zoo—	:	animal (Gr. *zoos*, animal)

The author solemnly promises that none of the words in the list will appear, henceforth, in this volume.

Here are the more common simple phobias with their common names: fear of animals, heights, closed spaces, doctors and dentists, wind, storms, lightning, loud noises, driving a car, flying in planes, travelling by underground railway, injections, and blood.

Fear of crowds is also common, but most often seen in agoraphobia, discussed in Chapter 5. A fear of public speaking is very common but comes under the category of 'social phobia' discussed in the next chapter.

Simple phobias

There are of course endless uncommon phobias, including fear of running water, of swallowing solid food, and of going to the hairdresser. There is even the case of the tennis player who wore gloves because he had a phobia about fuzz, and tennis balls are fuzzy. To repeat: phobias can develop toward almost any object or situation. However, some phobias are more common than others and apparently occur in every society. These are fears believed to be in part innate.

Children and animals are born with a fear of strangers and a fear of being looked at. This may explain why so many people are afraid of public speaking: the old inherited fears never quite go away and are often reinforced by experiences in later life. Fears of darkness, bats and other undomesticated creatures may be a carry-over of childhood fears of monsters lurking in the dark and may also be based on instinctive fear of objects and situations combining familiar and unfamiliar elements (Chapter 1). Dead people look *almost* like live people but not quite, and are frightening to children.

Even when phobias in adults have no apparent connection with instinctive fears of childhood, many at least seem understandable (more understandable anyway than the man who won't touch tennis balls). Injections hurt a little; dentists can hurt a lot. Nevertheless, when people who eagerly want to visit a foreign country stay home because they cannot stand the thought of having a smallpox shot, this is a phobia. When people walk around with rotting teeth because dentists terrify them, this is a phobia.

Animal phobias are the most common type of simple phobia. Two other common simple phobias are fear of heights and fear of illness. All three categories probably involve instinctual elements and all three are somewhat understandable in that animals can be dangerous, people do fall off high places, and illnesses are no fun. When these fears become exaggerated and unreasonable, however, they are distressing and even disabling. Each category deserves some additional comment.

Phobia: the facts

Animal phobias

These phobias have been studied more than any other. They almost always begin in childhood, often before the age of seven. They usually subside before puberty, but many adults continue to be bugged, so to speak, well into adulthood and often to the grave. Animal phobias are rarely seen by psychiatrists as a main complaint. How common they are in the general population is not known.

Most people with animal phobias are truly phobic of only one species. This may be cats, dogs, horses, other domesticated animals, birds, spiders, mice, worms, snakes, frogs, fish, and wild animals like bats. Many people are actively afraid of or at least squeamish about worms, mice, spiders, and snakes but the fearfulness is easily controlled and they do not actively avoid even the *possibility* of encountering one of these animals. Sometimes the phobia is directed toward one feature of the animal, such as feathers.

Impairment is associated with animal phobias when the animals cannot be avoided. For example, some city people have phobias toward pigeons and stay off the streets to avoid them (in which case the real diagnosis may be agoraphobia, as explained in Chapter 5). With respect to frogs, fish, and snakes, these can usually be avoided, sometimes at the expense of forgoing possible pleasure and even possible employment. Spiders? Avoidance is not so easy. The distress may be striking.

A woman with a fear of spiders screamed when she found a spider at home, ran away to find a neighbor to remove it, trembled in fear and had to keep the neighbor at her side for two hours before she could remain alone at home again; another patient with a spider phobia found herself on top of the 'fridge in the kitchen with no recollection of getting there — the fear engendered by sight of a spider had induced a brief period of amnesia. If in treatment sessions patients are brought too close to the spider, an acute panic ensues immediately in which they sweat and tremble and show all the features of terror which will wake them even from a deep hypnotic trance.

Simple phobias

This fear subsides when the spider is removed. (Reported by Isaac Marks.)

Animal phobias do not involve a fear of contamination by the animal from fleas, dirt, parasites, etc. People with exaggerated fears of contamination (from animals or other sources) almost always have obsessional neurosis (Chapter 7).

As said before, animal phobia almost always begins in childhood. Animal phobias starting later in adult life are more ominous, suggesting the occurrence of a major depression or other serious condition.

In other words, animal phobias have a characteristic 'age of risk'. If you do not have an animal phobia by the age of 12, you are unlikely to have one. (Most illnesses have an age of risk. If you are not schizophrenic by 25 or alcoholic by 45, it is unlikely you will become either. Having a characteristic age of risk is one of the reasons simple phobia was said in Chapter 2 to have a 'natural history'.)

Some other points about animal phobias:

1. Boys and girls are equally likely to have animal phobias until puberty, after which girls are more likely to have animal phobias (and women are more likely to have them later in life).
2. Unlike most psychiatric disorders, animal phobias do *not* run in families. Little girls with cat phobias as a rule do not have parents with cat phobias. This has two implications: (1) animal phobias are not directly influenced by heredity, and (2) children do not learn them from their parents. So, why do they occur?

Sometimes animal phobias seem related to a specific event. Father takes the children out to watch him drown the kittens, and at least one of them develops an intense desire to avoid kittens. A child gets bitten by a ferocious dog and thereafter avoids dogs, ferocious or otherwise.

However, most animal phobias, like most phobias in

Phobia: the facts

general, seem unrelated to stressful events. Even Sigmund Freud, one of the world's greatest explainers, had trouble explaining animal phobias: He wrote in 1913:

> The child suddenly begins to fear a certain animal species and to protect itself against seeing or touching any individual of this species. There results the clinical picture of an animal phobia, which is one of the most frequent among the psychoneurotic diseases of this age and perhaps the earliest form of such an ailment. The phobia is as a rule expressed towards animals for which the child has until then shown the liveliest interest, and has nothing to do with the individual animal. In cities, the choice of animals which can become the object of phobia is not great. They are horses, dogs, cats, more seldom birds, and strikingly often very small animals like bugs and butterflies. Sometimes animals which are known to the child only from picture books and fairy stories become objects of the senseless and inordinate anxiety which is manifest in these phobias. It is seldom possible to learn the manner in which such an unusual choice of anxiety has been brought about.

Of all phobias, people with animal phobia probably respond best to types of behaviour therapy described in Chapter 9, particularly desensitization.

Fear of heights

Fear of heights is really a fear of falling. It may be totally unrealistic. The person may be close enough to the ground for a fall not to hurt him, or be sufficiently protected by the environment for a fall to be impossible. No amount of reasonableness, however, curtails the fear. It is rooted in a basic instinct, the fear of receding edges discussed in Chapter 1 and shared by young goats and small children alike. Most children and goats lose the fear, but not all, and a very disabling fear it can be.

Some victims will not walk down a flight of stairs if they see the open stairwell. Some will not look out a window from the second floor or above, particularly if the window goes from floor to ceiling. Others will not cross a bridge on foot,

Simple phobias

although they may cross it by car without worrying.

Just as a fear of heights is really a fear of falling, a fear of falling is really a fear of loss of support. More exactly, it is fear of loss of *visual* support. Hence the person looking out a window experiences no fear if the window is waist-high. A car offers the same protection.

Behind the fear of losing visual support seems to be an even deeper fear: the fear of being *drawn* over the edge of the height. The high-up window, the car's doors, provide a sense of protection that no amount of reasoning-with-oneself can give.

The fear of losing support need not even involve heights, as the following case reported by Isaac Marks illustrates.

The patient was a 49-year-old housewife who as a child had been afraid of the dark. At the age of 48 while running for a bus she suddenly felt dizzy and had to hold on to a lamppost for support. This recurred and gradually the patient became unable to walk anywhere without holding on to a wall or furniture to support her — she was 'furniture bound'. Removal of visual support more than a foot away from her induced terror and crying. Actual physical contact with the supporting object was not essential. The patient was perfectly calm when sitting or lying down.

Developing the phobia at age 48 is unusual, particularly if this was her first phobia since the almost universal fear of darkness as a child. As will be discussed later, developing simple phobias for the first time in one's 30s or 40s suggests the presence of a depression or some other non-phobia disorder.

Even more unusual — or surprising, anyway — is a fear of heights rather frequently reported by airline pilots. As a consultant to the US Pilots Union, I have had a number of pilots tell me they panic looking over the top of buildings — but fly at 40 000 feet with nonchalance. Again, it is not the height that counts, but the 'visual space'. There is not much visual space in a plane, even in the cockpit.

Phobia: the facts

Illness phobias

We all have a fear of illness, though stoics may not admit it or show it. The lump, the cough, the spot on the skin: probably nothing, but who can be sure? Some people rush off to doctors at the first hint of trouble; most do not. They just worry for a while and hope it goes away. It usually does.

But a few continue worrying — even after the lump, the cough, the spot have disappeared. The fear becomes persistent, excessive, irrational: a phobia.

Illness phobias change with the times. For centuries, phobia of venereal disease was common. Generations of men (hardly ever women) inspected their genitalia for minute changes, and, finding none, still worried. They worried even if there had been no sexual exposure; virginal bachelors seemed particularly susceptible. There was always an explanation: toilet seats have been blamed as long as there have been toilets (and have always been as blameless). Doctors, finding nothing, shrugged or gave yesterday's equivalent of a vitamin B shot, which did not help phobic patients then any more than it does now.

Coughs are common, and so was tuberculosis phobia until a cure came along and people stopped talking about TB. Today, cancer phobia leads the list, with heart disease a close second. Growlings, gurglings, rumblings: all the normal cacophony of the intestinal tract invoke terror in the cancer phobic.

Illness phobics go to doctor after doctor, have examination after examination — and no matter how reassuring the former and negative the latter, are still not convinced. Excessive reassurance is particularly viewed with alarm.

Some illness phobics do the opposite: avoid doctors at all costs. They refuse to take insurance examinations because of what might be found. Suspecting a lump in a breast, they never touch the breast, even when bathing. If the disease they fear is ultimately found to exist, the fear often goes away. 'Fear is more pain than the pain it fears', Sir Philip Sidney said.

Simple phobias

Ironically, fear of illness may produce, if not illness, symptoms of illness. Fear both tightens and loosens the sphincter muscles, causing constipation, diarrhoea, or both in rotation. Fear makes the heart pound. Fear tightens scalp muscles, the real cause of headaches self-interpreted as incontrovertible evidence of brain tumour.

Fear, some believe, actually causes some illnesses such as asthma and colitis, but the evidence for this is weak. Fear undoubtedly worsens the symptoms or even precipitates attacks (this is not the same as causing an illness).

People with illness phobia often know friends and relatives who have had the feared illness, perhaps explaining the choice of the particular imagined illness but hardly its intensity and intractability.

Illness phobia needs to be distinguished from hypochondriasis. Hypochondriacs have many imagined symptoms and illnesses. They usually take their doctor's word that it is imaginary or 'psychological' or whatever euphemism he uses, only to return on another day with another imaginary illness.

Illness phobia fixes on one illness, and no amount of reassurance unfixes it.

More men than women have illness phobias; more women are hypochondriacs (or at least see doctors repeatedly for illnesses that they don't have).

Illness phobia also must be distinguished from delusions. Both phobias and delusions are fixed, false ideas. However, the delusional patient does not realize the idea is false. The phobic patient will grant that it is foolish, but be terrified anyway. Delusions occur in schizophrenia, manic-depressives, and brain diseases. Simple phobias are isolated symptoms in otherwise healthy individuals.

How common are illness phobias? Of simple phobias seen by psychiatrists (hardly a representative sample of the general population), 15 to 30 per cent are illness phobias.

Phobia: the facts

Natural history

As mentioned in Chapter 2, simple phobia has achieved the official status of 'disorder'. This means that it has been studied sufficiently for doctors to have some idea about when it begins and what happens: in short, a 'natural history' that helps physicians distinguish one disorder from another.

Simple phobias usually begin early in life. Animal phobias almost always begin before puberty and usually in early childhood. Other simple phobias may begin somewhat later but most begin at least by the mid or late twenties. An exception is illness phobia, which often begins in mid life.

If a simple phobia begins after the age of 30, it may not be a simple phobia but agoraphobia (Chapter 5) or a symptom of depression (Chapter 7).

Sometimes phobias develop so gradually that the victims have trouble remembering precisely when they began. Other victims recall vividly the onset of a phobia, particularly if it was accompanied by a panic attack. Many years later, they can identify the hour, day, and precise location where it happened. Vivid recall occurs less often with a simple phobia than with agoraphobia, since panic attacks occur more frequently in the latter condition.

Simple phobias occur in men and women about equally, with animal phobias occurring more in women and illness phobias more in men. More women are *treated* for phobias than men, but this may be because more women see doctors, regardless of the problem.

Once a phobia has lasted for a year, it tends to become chronic, meaning the fear may persist for many years unless two things happen: (1) the patient's life circumstances change so that he or she *must* confront the feared object or situation repeatedly, in which case the phobia has no 'choice' other than retreat in the face of necessity; (2) the patient receives treatment.

The first happens commonly. Trains are discontinued and

Simple phobias

the salesman with phobia about flying *must* travel by plane to keep his job. The recession worsens and the only jobs the young man can find is selling encyclopaedias, which he does in spite of his phobia about strangers. In other words, when the choice is avoidance or survival, the healthy person with an isolated phobia chooses survival — and after a time the fear goes away (although not always 100 per cent).

Confrontation of the feared object or situation is the foundation of the behaviour treatment of simple phobia (Chapter 9). The effectiveness of harsh reality (discontinuation of trains, recessions) in 'curing' phobias is undeniable.

In three studies, phobic individuals have been studied over a period of time to see what happens to them and their phobias. Here are the conclusions:

1. Phobias which begin before the age of 20 tend to improve and eventually disappear. Five years after the phobia begins, about half of the young victims are symptom-free and almost all are improved (less bothered).

Phobias which begin after adolescence continue for longer periods, with about half of patients improved after five years but only about 5 per cent symptom-free. In the older group, the phobia gradually becomes more severe in about one-third of patients. About 20 per cent are impaired in their work or social functioning because of the phobia.

Gender has little influence on the outcome of simple phobia: women do as well (or poorly) as men, with age of onset being the most important prognostic indicator.

2. Simple phobias as a group have a more favourable prognosis than the other phobic disorders. One reason is that simple phobias almost always involve a single isolated phobia whereas social phobias more often involve two or more phobias and agoraphobia invariably involves multiple phobias.

An exception is illness phobia. It typically occurs over the age of 40, when 'normal' fear of cancer and heart attacks is endemic (at least in the US). The late age of onset is untypical

Phobia: the facts

of simple phobias and carries a worse prognosis. The phobia, if anything, gets worse rather than better, and illness phobics are notoriously hard to treat. The illness phobic cannot avoid the feared object because there *is* no feared object. There is no possibility for the phobia to be cured by direct confrontation.

3. Patients with simple phobia (other than illness phobia) rarely see physicians for their problem. When they do, they respond better to treatment than other phobics.

4. Patients with simple phobias are no more likely to develop serious clinical depressions or other psychiatric disorders than individuals who have never had a phobia.

5. Patients with simple phobias come from reasonably stable families and have reasonably normal childhoods and marriages.

Some psychiatrists and writers on phobia take a more pessimistic view of simple phobia than I have presented here. The reason usually offered is that phobias tend to multiply, 'spread like the tentacles of a vine', as one writer put it.

John Bunyan wrote a book called *Grace abounding to the chief of sinners* (1666) with a passage which seems to support the pessimistic view. He tells how much he enjoyed ringing church bells and how the enjoyment offended his Puritan conscience:

> Now, you must know that before this I had taken much delight in ringing but, my conscience beginning to be tender, I thought such practice was but vain, and therefore forced myself to leave it, yet my mind hankered; wherefore I should go to the steeple house, and look on it, though I durst not ring. But I thought this did not become religion neither, yet I forced myself, and would look on still, but quickly after, I began to think: How, if one of the bells should fall? Then I chose to stand under a main beam that lay overthwart the steeple, from side to side, thinking there I might stand sure, but then I should think again. Should the bell fall with a swing, it might first hit the wall, and then rebounding upon me, might kill me for all this beam. This made me stand in the steeple door, and now, thought I, I am safe enough; for if a bell should then fall, I can slip out behind these thick walls, and so be preserved not withstanding.

Simple phobias

So, after this, I would yet to go see them ring, but would not go farther than the steeple door, but then it came into my head: How, if the steeple itself should fall? And this thought, it may fall for aught I know, when I stood and looked on, did continually so shake my mind, that I durst not stand at the steeple door any longer, but was forced to flee, for fear the steeple should fall upon my head.

In discussing Bunyan's plight, John Nemiah pointed out:

the tower of nearby Ely cathedral had, indeed, collapsed in the early 14th century, but in the intervening three centuries there could have been few if any others that had met a similar fate, and the statistical chances that the catastrophe he dreaded would occur were infinitesimally small. Bunyan demonstrated a further common feature of phobias — their tendency to spread. At first, it is just the straight plunge of the bell he feared, then its bouncing course and finally the complete crashing destruction of the whole steeple, so that he had to take ever and ever greater measures of avoidance to escape his doom.

This conclusion certainly applies to agoraphobia (Chapter 5). 'Spreading' is not a characteristic of simple phobia.

Simple phobia, in conclusion, is a rather benign disorder, particularly if it involves only a single phobia and begins in childhood or adolescence. Although the phobia may persist for many years, it rarely gets worse and often gets better and is not associated with more serious psychiatric disorders. The phobic anxiety is usually quite manageable and impairment slight, either because the dreaded object or situation can mostly be avoided or because, if unavoidable, the object or situation stops being dreaded.

4

Social phobias

> Fear of two staring eyes is ubiquitous throughout the animal kingdom
>
> Isaac Marks

A social phobia is basically a fear of being looked at. The fear may be partly innate (see Chapter 1). Monkeys and other animals dislike stares. But when stares, real or imagined, produce extreme discomfort in *particular situations*, the result is a social phobia: 'social' because scrutiny by other people is always in some way involved.

Behind the fear of scrutiny is a fear of being embarrassed, ridiculed, or making a fool of oneself. And behind this fear is a *performance fear* — the fear of being unable to perform or fear of losing control in some way when others are looking on.

Performance fear often produces what is most feared: poor performance. The panic-stricken public speaker cannot utter a word; the person afraid to eat in public gags at the sight of food.

In some instances the social phobia is the fear of being observed to have a social phobia: a telltale blush or tremor.

Avoidance of the feared situation — the *sine qua non* of any phobia — usually results in inconvenience and distress but rarely incapacitation.

Here are some social phobias, starting with the most common:

Fear of public speaking

A little stage fright is considered useful. Many professional performers — lecturers, singers, actors — say they have to be

Social phobias

a little nervous to be at their best. But extreme stage fright is another matter. It leads to avoidance of stages. Many people can go all their lives without the risk of having to give a talk in public, but often at a cost. The talented singer never sings, the born teacher never teaches. The salesman never becomes a sales supervisor because he would be expected to speak before groups. In its most extreme, fear of public speaking may lead parents to avoid PTA meetings or Sunday school. They might be called upon to say something.

Fear of using public lavatories

Performing bodily functions (as the euphemism goes) is not as simple as it might seem. It calls for some intricately timed and synchronized tightening of some muscles and loosening of others. Moreover, performing appropriately — i.e. in the right location at the right time — is highly valued in sanitation-conscious middle-class Western civilization (more so than in older, perhaps just-as-civilized parts of the world).

One of our first public performances involves the act of elimination. Successful performance is rewarded with smiles and congratulations. The soiled diaper, on the other hand, elicits a reaction the child only learns the name for later: disgust.

Presumably children are not born with disgust toward their bodily products, but the force of the mother's response assures the early acquisition of this interesting response to such things. By age three or four, most children are as disgusted by faeces as their mothers are, and have discovered bad smells from other sources, such as vomit. (Negative reactions to certain odours may have survival value and therefore a genetic basis; spoiled food is as noxious to many animals as to people. Our culture, however, has extended the range and subtlety of offensive odours to an unprecedented degree, creating a huge market for deodorants.)

Freudians make much of toilet training as a source of later

Phobia: the facts

adult ambivalence: whether to go or not to go. Indecisive people are supposed to have had unfortunate toilet training experiences. Some Freudians also believe the act of 'going' is early identified by the child as aggressive and therefore potentially useful in combating a hostile environment. The association of excretory processes and aggression seems borne out by the choice of some well-known four-letter words as terms of abuse. According to the anthropologist Margaret Mead, every society has a stock of words based on excrement and sexual acts to express anger.

Another Freudian notion is that the act of urination, at least by men, becomes confused at an early age with competition with one's peers. It is true that boys will sometimes compete to see who can urinate the furthest. One twist of this quaint sport (if mothers only knew!) consists of writing words in the snow with urine tracks, with some boys showing more skill than others.

For most young boys this is a 'phase' which they pass through quickly, but a few become obsessed with competition on this primitive level, resulting in what Freudians call a 'Urethral Personality' (later renamed Type A personality).

Toilet training, disgust, ambivalence, aggressiveness, competitiveness: what do these have to do with social phobias? Perhaps nothing, but, then, there is always the problem of explaining why some people develop a particular phobia and most do not. In the case of toilet phobias, the Freudian ideas may not explain the whole story but are at least plausible.

A 20-year-old Army inductee sought out a base psychiatrist with the following problem. Since the age of 13 or 14 he had become increasingly apprehensive about using the toilet if there was any possibility someone might see him. Even when he had the urge to defaecate, he would wait until no one was in the house or it was late at night and others were asleep. Even the thought that someone might *know* he was using the bathroom made him fearful. When fearful, his sphincter muscle tightened and he was unable to have a movement.

Social phobias

He went out of his way to discover . . . isolated places where he could use the toilet. Finding such places became a major preoccupation. Nevertheless, he was able to attend school and never told anyone about his problem. The sound of people talking elsewhere in the building would have the same effect as someone entering the lavatory: he would panic, his sphincter would tighten, and he could not perform.

The prospect of entering the army produced extreme apprehension. There is nothing less private than an army barracks. Nevertheless, after a few days of constipation, he found unused barracks with abandoned facilities. He set the alarm for 3.00 a.m. in the hope of finding his own barracks lavatory unoccupied. Then he received orders that he was being shipped overseas. This produced a tremendous crisis and he sought help from the psychiatrist, who got the orders rescinded.

Later in life (the psychiatrist learned) the patient married and had the relative privacy of a home bathroom. His phobia gradually diminished and, except for some residual anticipatory fear when invited overnight to someone's house or other unpredictable situations, the phobia disappeared.

Was the phobia related to toilet training? According to this patient, his mother was a super vigilant toilet trainer. Every day she would ask her child, 'Have you gone?' If he had not, she plied him with laxatives. Missing *two* days was a national crisis, calling for enemas. Three days and the doctor was called.

Possibly, years later, scrutiny by *anyone* came to represent scrutiny by the mother, armed with her punitive laxatives and enemas in the event of a flawed performance. It is possible, but hard to prove. Few social phobias lend themselves to such a handy explanation.

How common are social phobias? Nobody knows, but any army sergeant in a training camp will tell you that constipation is common during the first few days of an inductee's service. Whether lack of toilet privacy is the reason is not known.

Perhaps a more common toilet fear involves apprehension about urinating in public. Many men become anxious standing at a urinal with a line of men waiting behind them.

Phobia: the facts

Their sphincters tighten and they become embarrassed by what seems the long delay. Sometimes they depart the urinal unrelieved to avoid attracting more attention.

One reason to think this phobia is fairly common is a scene often observed in large male lavatories at sporting events. Long lines form behind the closed booths despite the fact that urination is the only goal and there are open spaces along the row of urinals. One interpretation is that a sizeable number of men have at least a mild social phobia about the act of urination.

What about women? Again, no one knows the frequency of elimination phobias in women, but gynecologists know of women who bring a urine specimen with them for an office visit in case one is needed, fearing they cannot 'go on demand' in the doctor's toilet.

The competition theory may apply to men, fearful of appearing inferior to their peers, but what about women? Success for little girls presumably is not measured in terms of the range and arc of a flow of urine. The story is incomplete — as are all theories about phobias.

Fear of eating in public

An attractive 30-year-old wife of a business executive refused to entertain clients and associates of her husband with a dinner party or attend dinner parties in restaurants or someone else's home. Her husband made excuses for her but felt his career was being damaged by their lack of social life. At first her explanation for refusing to join in dinner parties was that it was a waste of time or too much trouble. Later she confessed, tearfully and with much embarrassment, that she was afraid of being *unable* to eat if strangers were watching. Questioned by her concerned husband, she said she was afraid that once food was in her mouth she would be unable to swallow it and then have the embarrassing situation of not knowing where to dispose of the food. She was also afraid that she would gag on the food and possibly vomit.

The phobia began in her late teens when she found the sound of other people eating offensive. Then she became afraid that her own

Social phobias

chewing and swallowing would offend others and made a great effort to chew and swallow quietly. This developed into a fear of swallowing any food at all and later a concern that if she did swallow she would vomit, for her the ultimate in shameful acts. She could eat in front of her husband with no difficulty until she told him about the phobia and then became concerned that *he* was watching her eat and expecting her to gag or vomit. From then on she ate alone in the kitchen. The marriage survived, but barely.

Fear of eating in public does not involve loss of appetite. Weight loss may occur but not because the person wants to lose weight or has an obsession with being thin, as happens in anorexia nervosa. In a sense, patients with anorexia nervosa have a severe food phobia, but anorexia nervosa is accompanied by physiological changes such as cessation of menses and bizarre behaviour such as hoarding food and self-inducing vomiting.

Sometimes fear of eating and drinking in front of others is related to concern that one's hands may tremble while bringing the fork or cup to mouth. The fear of dropping food or spilling coffee in truth is a performance fear, a desire to avoid calling attention to a perceived weakness.

Eating phobias in one way resemble toilet phobias. Both involve semiautomatic functions involving reflexes and muscles of the alimentary canal. These reflexes and muscles are sensitive to the effects of strong emotion such as fear and anger. Fear and anger may cause gagging and tighten the anal sphincter muscle. When a social phobia involves fear of performance involving these delicate mechanisms, the fear itself increases the chance of non-performance.

Fear of vomiting

This may exist independent of an eating phobia. Isaac Marks tells of a 34-year-old unmarried secretary with a fear of vomiting of 13 years' duration.

Although the patient remembered being concerned at other children vomiting when she was 5 years old, she did not develop the phobia until 21, when she became afraid that other people or she herself would vomit on the train and she began avoiding traveling situations. The fear became fluctuatingly worse in the next five years. The patient would awake at 5:15 a.m. daily in order to travel to her office before the rush hour. She avoided eating in public places, in restaurants, or in strangers' homes. She also avoided going to theaters with friends because it was easier to leave the theater if she was alone in the event a fear of vomiting developed.

In fact, the patient had never vomited in public and had not seen anybody else vomit for many years.

Away from the phobic situation the patient had no anxiety and was not depressed. Her work was satisfactory. The patient had four years of psychotherapy, but her phobia remained unchanged. Later, after 12 sessions of desensitization [see Chapter 9], the patient became able to eat out in restaurants and travel in crowded trains.

Fear of sexual performance

Sexual malfunctioning is not usually considered a form of phobia, but, in fact, certain forms of sexual malfunctioning fit the concept of social phobia. The person is anxious about performing because someone is observing his or her performance; the performance anxiety itself increases the chance of non-performance.

Premature ejaculation, impotence in the male, and frigidity in the female, are the most common forms of social-sexual phobia. Numerous books have been devoted to the subject and it will not be explored here other than to say that the treatment for a social-sexual phobia resembles the treatment for other types of phobia. By various means, performance anxiety is gradually attenuated so that performance can occur.

This type of phobia also involves the essential feature of any phobia, namely, avoidance of the feared situation. Fear of sexual performance probably is one of the world's most effective forms of birth control.

Social phobias

Fear of being watched at work

How many times have you heard: 'Don't look over my shoulder while I'm working'? It happens often enough to suggest this is a common social phobia.

It is truly a phobia only if it leads to some incapacity. If the secretary cannot type or the computer operator cannot punch keys if someone is watching (or may be watching), this reduces output and threatens job security. The phobia can apply to any mechanical operation: taking shorthand, writing on a blackboard, sewing, knitting, or even buttoning a coat if somebody is watching. Teachers need to write on blackboards and seamstresses need to sew. Social phobias are always an inconvenience but sometimes the consequences are very serious indeed.

Some people become phobic about writing or handling money in front of others. For example, a person might wait outside a bank until it is empty so he can deposit money. Such fears are often related to a fear of trembling.

'Writer's block' may be phobic, or may reflect some other problem (depression, lack of motivation). The graduate student who has completed his work toward a degree and cannot face writing his thesis may have a social phobia; what he cannot face is scrutiny.

Fear of crowds

The person with a social phobia of crowds must be distinguished from a person with agoraphobia (Chapter 5). Social phobia of crowds involves a fear of being watched. Fear of crowds in agoraphobia involves a fear of being enclosed or suffocated. The latter is far more serious than a social phobia, where the person can usually engage in almost any activity as long as nobody is watching. There may be inconvenience. For example, crowd phobia may restrict a person's activities so that he only goes shopping at hours when few people are

Phobia: the facts

around and avoids sporting events. It may involve a fear of eye contact; he may stand up in the train to avoid eye contact with sitting passengers.

However, a social phobia is usually not crippling assuming there is only a single phobia. Agoraphobia involves multiple phobias and is crippling indeed. Hippocrates had a patient with a crowd phobia:

> Through bashfulness, suspicion, and timorousness, he will not be seen abroad; loves darkness and cannot endure the light; his hat still in his eyes, he will neither see, nor be seen. He dare not come in company for fear he should be misused, disgraced, overshoot himself in gesture or speeches, or be sick; he thinks every man observes him . . .

Of course, this may be paranoid schizophrenia. If the patient knows his fear is absurd, it is a phobia; if he considers it justified, it is a delusion and a symptom of schizophrenia or some other psychosis.

Social phobias are often accompanied by a fear of fainting, stumbling, dropping things, or otherwise attracting attention. Phobic people rarely do these things. The only phobia really associated with fainting is a phobia about blood. For some reason, the sight of blood sometimes slows the heart, resulting in a faint (many doctors have seen it happen).

Fear of blushing

This is a particularly painful and stubborn phobia, described by the French neurologist Pierre Janet in the nineteenth century. According to John Nemiah:

> The patient, usually a woman, is terrified that she will blush in the company of others and is convinced that in that state she will be highly visible and consequently the center of painful attention. If questioned the patient cannot say what is so dreadful about blushing, but it is often evident that shame is an important component of her anxiety. A change in color may not be at all evident to the observer, despite the fact that the patient insists that she feels bright

Social phobias

red; the force of her fear, unfounded as it may be, often leads the patient to a severe restriction of her social life.

Fear of being touched

Just as some people avoid being watched in certain situations, others avoid being touched.

A 23-year-old jazz pianist could tolerate being touched by his girlfriend but by no one else. He played the saxophone as well as piano, but limited his playing to the piano where he would be less likely to be touched. He never played with a travelling group because touching was unavoidable with others in a car or van. He stayed out of elevators, not because of fear of closed places but because of fear of being touched.

He could not explain why touching bothered him. He was not concerned about dirt or contamination from others. He did not associate the fear with sexual contact. He just hated being touched and felt that way since 15 or 16 when the fear started for no apparent reason.

Once he declined a ride to a sauna to get a massage when he learned there would be other passengers in the car. He looked forward to the massage but the thought of brushing someone's shoulder on the way was intolerable.

Reciprocal inhibition — the psychological principle which holds that one cannot experience two opposing emotions simultaneously — may explain why the pleasure from his girl friend's touch overrode the incompatible fear of being touched (Chapter 9).

Miscellaneous social phobias:

A fear of practising musical instruments because the neighbours will hear mistakes; a fear of swimming or undressing in front of others because of shame of one's appearance; a fear of driving automobiles; a fear of criticism from superiors (the real reason why some people refuse to work for anyone but themselves).

Phobia: the facts

Most social phobias first occur in adolescence or the early twenties. They are rare before puberty or after 30. Social phobias appear to affect both sexes about equally. There is some evidence that social phobia is more associated with depression and other psychiatric problems than is simple phobia. Most people seek professional help for a social phobia because of a complication of the phobia rather than the phobia itself. In one series of patients, 18 per cent had an alcohol problem. This and depressions are the commonest reasons for social phobias to be referred to a psychiatrist.

Most social phobias develop over several months, stabilizing over a period of years with a gradual diminution of severity in middle life. As children, people with social phobia are sometimes described as 'sensitive', but most sensitive children do not develop social phobias.

Most social phobias begin without any apparent precipitating event, such as loss of a family member or emotional trauma. Social phobias do not run in families, so there is no evidence for a hereditary factor. In truth, social phobias are a mystery. They have been little studied and their cause is unknown.

5

Agoraphobia

> When a trout rising to a fly gets hooked on a line and finds himself unable to swim about freely, he begins a fight which results in struggles and splashes and sometimes an escape. Often, of course, the situation is too tough for him.
> In the same way the human being struggles with his environment and with the hooks that catch him. Sometimes he masters his difficulties; sometimes they are too much for him. His struggles are all that the world sees and it usually misunderstands them. It is hard for a free fish to understand what is happening to a hooked one.
> Karl Menninger

In 1871 a German neurologist named Karl Westphal described three male patients who shared a common symptom: all three became exceedingly anxious walking through an empty street or crossing an open space. He called the condition *agoraphobia*, '*agora*' being the Greek root for market place or place of assembly.

The men also had another symptom in common — a dread of crowded places — but Westphal made less of this. They all had found ways of relieving their anxiety, such as carrying a cane or umbrella, having a few drinks, being with a trusted companion.

The men had the usual physical symptoms of anxiety: palpitations, trembling, feeling of warmth, dry mouth, sweating, and breathlessness. They particularly complained of dizziness. And, in fact, some clinicians felt the dizziness was more important than the anxiety and called the condition 'dizziness in public places', perhaps caused by a disorder of the eye muscles.

The term agoraphobia has now survived for a century and at

Phobia: the facts

least is easier to pronounce than *Platzschwindel,* the German word for dizziness in public places. Moreover, it is now recognized that dizziness is a minor symptom. Sometimes even a fear of open spaces is not present. Fear of crowded public places today is believed to be the single most common complaint by sufferers of agoraphobia.

But agoraphobia *never* refers to a single complaint. It refers to a cluster of complaints. The fear of crowded places and, somewhat less often, fear of open spaces are the main complaints but the patient with agoraphobia also dreads some or all of the following:

(1) *Public transportation*: trains, buses, underground railways, planes. When crowded, these vehicles become intolerable. Waiting in a queue is almost as bad, whether for a bus or a cinema.

(2) *Other confined places*: tunnels, bridges, elevators, the hairdresser's chair, the dentist's chair, and the barber's chair. (Agoraphobia has been called the 'barber's chair syndrome'.) These fears belong to the category of 'claustrophobia', but most people with claustrophobia have only a single phobia and are not agoraphobics.

(3) *Being home alone.* Some agoraphobics require constant companionship, to the despair of friends, neighbours, and family.

(4) *Being far away from home* or in places where help cannot be readily obtained if needed. The agoraphobic is sometimes comforted just knowing there is a policeman or a doctor somewhere nearby.

The agoraphobic usually has difficulty explaining *why* they are afraid. Many say they are afraid of fainting, having a heart attack, dying among strangers. They fear becoming insane. They fear losing control in some manner; screaming or attacking someone (perhaps sexually) or otherwise attracting unwanted attention. The fears are groundless, and they usually know it. People with illnesses which may cause

Agoraphobia

fainting, heart attacks, or death almost never have agoraphobia. Agoraphobics do not become insane and do not lose control of themselves in public places.

Agoraphobics have multiple phobias, usually not the case in individuals with simple or social phobias. In addition, agoraphobia involves more features than simple or social phobias. Even when not in clearly defined situations that almost always cause intense fear, agoraphobics still tend to be anxious a good deal of the time. They are subject to depression, especially when thinking about their fears and how their life has been affected by them. In fact, depression is sufficiently common and severe that some authorities believe agoraphobia is a form of depression.

Agoraphobics frequently experience an eerie feeling called 'depersonalization'. The feeling is hard to define. In fact, if you describe depersonalization to someone who has never experienced the feeling, they usually shrug and look puzzled. People who have experienced depersonalization know exactly what you are talking about and describe the feeling as scary and very unpleasant. It involves feeling unreal, strange, and disembodied, cut off from one's surroundings. Patients with epilepsy involving the temporal lobes often describe the same feeling, and this has led to speculation that agoraphobia may be a form of epilepsy. The evidence for this, however, is almost nonexistent.

Agoraphobics are often inhibited sexually. One of the side benefits of recovery can be great improvement in the person's sexual life.

Agoraphobics are often described as having a particular type of personality. According to some observers, they tend to be passive, highly dependent on others, and give the impression of being in a constant state of alertness.

Agoraphobia is the most disabling of the phobic disorders and the hardest to treat. When severe, it can be as disabling as the most crippling forms of schizophrenia. Although agoraphobics almost never require chronic hospitalization, they are

Phobia: the facts

sometimes unable to leave home for months or years at a time. The more extreme cases may confine themselves to a single room or spend most of their time in bed. The term 'housebound housewife' usually refers to agoraphobics.

Even milder cases involve restrictions in social functioning. Victims are unable to visit friends and neighbours, or go on family outings. They recruit other people to do their shopping and take their children to school. They postpone seeing the dentist and may cut their hair themselves.

Surprisingly, family life is not as damaged by this as one might expect. Spouses will complain about their agoraphobic mate but their marriages apparently survive as well as most. The children tend to be sympathetic and do everything they can to help. The mental health of the children does not seem to suffer. There is one study suggesting that children of agoraphobics more often have school phobia than usually happens, but most studies indicate the children are normal.

Agoraphobics are amazingly inventive in discovering ways to mitigate their anxiety so they can maintain at least some social functioning. As Westphal reported, sometimes just carrying an umbrella helps. Other inanimate objects reported to provide relief include canes, shopping baskets on wheels, a bicycle pushed down the street, a folded newspaper carried under the arm. One agoraphobic found relief by loosening his belt. Another said that sucking a sweet gave comfort. Agoraphobics almost always are more comfortable in dark places than in sunlight. Some wear dark glasses. They feel better when it is raining. Others can confront crowds if they have a bottle of ammonia or tranquillizers in their pockets (never used). If they go to theatres at all, they find an aisle seat near the back to make a fast getaway if necessary.

Studies by Isaac Marks and his colleagues at The Maudsley Hospital in London have uncovered these and many other strategies agoraphobics have discovered to relieve their fears. To the above list Marks adds the following:

Agoraphobia

Deserted streets and vehicles are much preferred. Trains are easier to go on if they stop frequently at stations, and if they have a corridor and a toilet. Some journeys are easier if they pass the home of a friend, or a doctor, or a police station, when the patient feels that help is at hand if they get panic-stricken. In such instances if patients know the friend or doctor is not at home their journey becomes more difficult. It is the *possibility* of aid which helps them in their acute anticipatory anxiety before the journey. One patient was able to go on a particular bus route because it passed a police station outside which she would sit if the tension got too much for her. Agoraphobics usually find it easier to travel by car than any other means and may comfortably drive themselves many miles even though they cannot stay on a bus for one stop. Patients with cars may be able to hide their problem for many years.

Everyone agrees, however, that the most reliable fear-reducer is a trusted companion. Many agoraphobics only venture out of the house when accompanied by someone they know and trust: a husband, friend, child, or even a dog.

According to Marks, a popular writer compiled many of these features into a composite figure called Aggie Phobie: a woman walking at night up a dark alley in the rain while wearing dark glasses, sucking sweets vigorously in her mouth, with one hand holding a dog on a leash, the other trundling a shopping basket on wheels.

The symptoms of agoraphobia tend to wax and wane over the years, waning in part because of the victims' heroic struggle against their fears. Nowhere is this better described than in the following account by a woman writing in the *Lancet*:

For three years I had been unable to make a train journey alone. I now felt it was essential to my self-esteem to do so successfully. I arranged the journey carefully from one place of safety to another, had all my terrors beforehand, and travelled as if under light anesthesia. I cannot say I lost my fears as a result, but I realized I could do what I had been unable to do.

Now, like others who are disabled, I have my methods. The essentials are my few safety depots — people or places. The safety radius from them grows longer and longer. I am still claustrophobic; that

Phobia: the facts

rules out underground trains for me, and I use the district railway. I find it difficult to meet relations and childhood friends, and to visit places where I lived or worked when I was very ill. But I have learnt to make short visits to give me a sense of achievement and to follow them when I am ready for it by a longer visit. Both people and places are shrinking to their normal size. Depression usually returns about a week before menstruation, and I have learned to remind myself that life will look different when my period begins . . . I am also learning that it is permissible to admit to anxiety about things I have always sternly told myself are trifles to be ignored. Many of them, I find, are common fears.

If I am fearful of going anywhere strange to meet my friends I invite them home instead, or meet them at a familiar restaurant . . . Strangers, too, can be more helpful than they know, and I have used them deliberately; a cheerful bus conductor, a kindly shop assistant, can help me to calm a mounting panic and bring the world into focus again. If I have something difficult to do — to make a journey alone, to sit trapped under the drier in a hairdresser's, or to make a public speech — I know I shall be depressed and acutely afraid beforehand. When the time comes I fortify myself by recalling my past victories, remind myself that I can only die once and that it probably won't be so bad as this. The actual experience now is not much worse than severe stage fright and if someone sees me to the wings I totter on. Surprisingly, no one seems to notice . . .

I dare not accept my sickness — fear — because it never stays arrested. My very safety devices become distorted and grow into symptoms themselves. I must therefore, as I go along, break down the aids I built up; otherwise the habits of response to fear, or avoidance of occasions of fear, can be as inhibiting as the fear itself.

'It is hard for a free fish to understand what is happening to a hooked one', writes Karl Menninger. Clearly, we may not even know when someone is hooked. How many umbrellas are safety devices — and not against the weather? What are sunglasses shading? Sunlight or fear? How often is the drunk on the street there only because he is *drunk? Nobody knows how many agoraphobics are among us in various disguises, but the number may be larger than anyone suspects.*

Westphal in his 1871 paper mentioned that alcohol emboldened his three male agoraphobics to venture out into

Agoraphobia

open spaces. Other agoraphobics have discovered the same thing, and that tranquillizers also help. Are agoraphobics susceptible to alcoholism and drug abuse?

The evidence is conflicting. In one study, two-thirds of a group of alcoholics gave a history of agoraphobia or social phobia. (There is no comparable study of agoraphobia among drug addicts.) A study of agoraphobics, on the other hand, showed a relatively low rate of alcoholism and drug addiction (5–10 per cent). Since alcohol unquestionably reduces anxiety, it certainly would 'make sense' if anxious people, including agoraphobics (probably the most anxious of all), resorted to the grain or the grape for relief. It would make further sense that, having resorted to alcohol on repeated occasions, dependence on alcohol would occur.

In fact, most long-term studies do not indicate that alcoholics were unusually anxious as children or adolescents. However, phobia, as a distinct clinical disorder, must be distinguished from anxiety, and has been little studied in connection with alcoholism. Other studies are needed.

More consideration has been given to the possibility that agoraphobia is related to clinical depressions. In common with clinical depressions, agoraphobia often begins 'out of the blue': the first symptoms occur for no apparent reason.

Also, agoraphobics often are depressed and their depressions may resemble the depression seen, for example, in manic–depressive disease. Conversely, symptoms associated with agoraphobia are not uncommon in individuals with depression. Therefore, the issue of whether agoraphobia is a form of depression has not been resolved. A recent study suggested they are not the same illness. A laboratory test called the dexamethasone suppression test which is sometimes useful in diagnosing depression has been reported negative in agoraphobia.

Agoraphobia does not always begin out of the blue. Sometimes it is preceded by a physical illness or stress such as an important examination or marital conflict. The symptoms may come on all at once or gradually. Some agoraphobics have no symptoms at all until one day they are standing at a bus stop or shopping in a large store when they suddenly panic, rush home, stay indoors, and for years later avoid a variety of public places, including bus stops and large stores. Other agoraphobics are not able to say exactly when their illness began.

Once the illness has developed, victims often have trouble deciding which is worse: the anxiety that occurs in the situation itself (a store, a bus) or the anticipation of becoming anxious. The agoraphobic may start feeling anxious the moment he awakens, thinking about the day ahead. Some find the anticipatory anxiety is greater than the anxiety in the feared situation — an observation with treatment implications (Chapter 9).

Only one study indicates that agoraphobia may run in families. Other studies show high rates of alcoholism, depression, and various anxiety disorders in the families of agoraphobics, with female relatives more likely to be depressed or anxious and male relatives more likely to be alcoholic. This leaves the issue of possible hereditary factors unresolved.

How common is agoraphobia? One estimate is that it occurs in about one in every 160 adults. Most estimates of psychiatric illness are based upon clinic statistics. These figures invariably are skewed, since factors other than the illness enter into whether a person seeks treatment: income, availability of clinics, etc. With recent publicity about agoraphobia, more and more patients are becoming visible in one way or another. Little attention was paid to the disorder until recent years — one reason perhaps being that phobics, and especially agoraphobics, tend to be secretive about their symptoms. As for physicians, many have been unaware that the disorder existed.

When patients do see doctors, it is usually several years or longer after the illness has begun. The likelihood of seeing a doctor apparently does not depend upon the severity of the illness but on the person's ability to confide in others. Some

Agoraphobia

agoraphobics do not see doctors because they are too housebound to venture out to the doctor's office. Psychiatrists, who presumably would see more agoraphobics than other physicians, report that only about 2 per cent of their patients have this disorder.

Agoraphobia usually begins in the mid or late twenties, almost never occurring before 18 or after 35. Westphal's three patients were men, but at least two-thirds of agoraphobics today are women. Of the three types of phobia, agoraphobia has the latest age of onset, with simple phobia occurring in childhood and social phobia in adolescence.

What happens to agoraphobics? If they do not receive treatment, about one-half to three-quarters of the mild or moderately disabled agoraphobics will recover or substantially improve five to ten years after the illness begins. The severely disabled housebound patient may remain that way indefinitely. Some cases are short-lived but nobody knows how many. Most clinicians would agree with Isaac Marks that if the symptoms of agoraphobia persist for a year or longer they are almost certain to last much longer.

New forms of treatment for agoraphobia have been developed in the past few years and look promising. Chapter 9 describes them in detail.

6

Phobias in childhood

> Men fear Death, as children fear to go in the dark; and as that natural fear in children is increased with tales, so is the other.
>
> Sir Francis Bacon

A phobia is an intense, recurrent, unreasonable fear. It is a specific fear, directed toward a person, animal, object, or situation. It leads to avoidance of whatever is feared.

That is how phobia was defined in Chapter 2, referring mainly to adults. It is harder to apply to children.

For one thing, in small children, you have to infer feelings from behaviour. Children are *intense* about most things — likes and dislikes. Growing up is a gradual process of de-intensification.

What is *unreasonable* is often a matter of opinion. Lightning does strike, snakes do bite. And how does a small child know what is out there in the dark? How does he know the encyclopedia salesman is not the Devil? Viewed as a natural product of inexperience, most childhood fears do not seem unreasonable at all.

Then, are they phobias? We'll call them that if they seem excessive or handicapping.

The most common fear in childhood is a fear of animals. This usually comes on between the ages of two and four and is gone before age ten. The next most common fear is of darkness, a fear experienced by children between the ages of four and six. It is not really the darkness itself, but what is out there *lurking* in the darkness that seems to be the source of fear. Most young children are afraid of storms, thunder, and

Phobias in childhood

lightning and stop being as afraid as they get into their early school years.

Most children are afraid of doctors. Is this a phobia? If the child has pneumonia and runs away from home to avoid seeing a doctor, it is.

About one-third of two-year-olds are afraid of strangers, running and hiding at the sight of one. The fear appears short-lived. Fewer than 10 per cent of four-year-olds are afraid of strangers.

Almost all children are afraid to be left alone for very long before the age of five but few children mind it when they are older, and when they are *really* older, they often prefer to be alone (particularly if 'being alone' means bringing home other teenagers).

Snake-fear is full-blown in about one-third of children by the age of two and, unlike the fears mentioned above, tends to hang on. Snake phobia is probably the most common phobia in adults. Perhaps if snakes were as common as darkness and strangers this fear too would disappear.

Little children hardly ever fear spiders. In adults it is common. Why? Nobody knows.

There are other less common fears too numerous to mention, but having a fear of some kind that seems excessive and unreasonable, at least to the adults around, occurs in perhaps 90 per cent of children (and, like masturbation, those who deny it are probably lying).

Parents tend to underestimate the number and intensity of the fears their children experience, judging by studies where children were taken aside and really interrogated about their fears. It probably does not matter much. Very few children have fears sufficiently serious to warrant concern. An infinitesimal number get taken to doctors because of fears. Except in the case of school phobia, child psychiatrists hardly ever see a child because of phobias.

Apparently more girls than boys have phobias, perhaps because boys learn from the cradle that it is not manly to be

Phobia: the facts

afraid. One study reported that blacks have more phobias than whites, although why this should be defies speculation.

At the time of adolescence most children have outgrown their fears, or at least they are not so noticeable. There are two phobias which peak in adolescence: fear of blushing and fear of being looked at. They peak in girls before they peak in boys, perhaps because girls hurtle into adolescence before boys. (The maximal height increment occurs at 10 years in girls and at 12 years in boys. The maximal weight increment occurs at 11 years in girls and at 13 years in boys. There are also other differences between girls and boys, which both sexes have usually detected by this time.)

It is easy to say that children are afflicted with unreasonable fears because they are unreasonable little people and, as they learn more, they shed the unreasonable fears of childhood and assume the reasonable fears of adults. Still, this does not explain why some children have more fears than others and why one child is terrified of such-and-such and another is unaffected.

One can always say it is 'constitutional'. No doubt it is, in part. Take the earliest 'fear' response of all: the startle response to a loud noise. Infants vary widely in their startlability. One child will jump out of his crib at the sound of the wind and another will respond to a cannon shot by turning over and going back to sleep. It is impossible to ascribe these differences to mothering or anything else going on in a baby's life.

Are infants who startle easily more prone to phobias later on than their less flappable crib mates? There is no evidence to prove it. Childhood phobias are not correlated with nightmares, bedwetting, or psychiatric illness in later life. Perhaps the reverse: hyperactive children are said to have fewer fears than most.

Phobias in childhood (as in adults) sometimes begin with a traumatic experience. A child is in a car accident and for a long

Phobias in childhood

time refuses to ride in cars. He walks out too far in the lake, his head goes under, and never goes back in lakes again.

Phobias sometimes develop after a tonsilectomy, the death of a relative or classmate, or even a class in sex education. The connection between the trauma and phobic behaviour may be unobvious to the outsider: avoidance of relatives of the deceased, unwillingness to use public toilets.

There is one report correlating birth complications with phobias in childhood. One report does not make it true, but it should be considered as a possibility.

Children are less fearful in the presence of a comforting mother, father, older sibling, or trusted adult. Children who lack such people also lack the comfort they provide and presumably are more subject to phobia. That would make sense anyway. As a matter of fact, there is little evidence that children who come from broken homes, or no homes, or are raised in orphanages or concentration camps are more afflicted by 'unreasonable' fears than are children in more secure circumstances. They have plenty of reasonable fears (fear of neglect, desertion, and starvation are eminently reasonable) but there is no reason to believe they are particularly prone to unreasonable or 'neurotic' fears, possibly because the reasonable fears leave them little energy for the unreasonable.

Paradoxically, overprotective mothers are said to have especially fearful children.

At any rate, trauma and missing parents cannot be held accountable for most childhood fears. They simply come and go. Yesterday your two-year-old ignored the roar of the vacuum cleaner. Today he is terrified — and there has been no intervening experience with vacuum cleaners or other loud noises to explain it. 'We can only infer', according to one scholar, 'that the child's perceptual response to the vacuum cleaner is undergoing change as a result of the joint interaction of maturation and learning.' Whatever that means.

Phobia: the facts

This chapter would not be complete without mention of what might be called Freudian Fears.

Here are three Freudian Fears:

Tiny babies, nursing at the breast, are afraid of biting the nipple and thereby losing their food source. A little later they are afraid of defaecating because a piece of themselves disappears down the toilet and also because it is perceived as an act of aggression towards the much-needed mother.

Around four or five, little boys become afraid they will lose their penis. There are three reasons. First, they tend to enjoy their penis. Second, they have noticed that little girls don't have penises. They assume girls *once* had penises but lost them because of a misdeed. Third, they conclude that the misdeed leading to loss of penises is lust (in the case of a little boy) for his mother which incites the big rival-father into taking punitive action: castration of the son.

These fears are believed to cause terrible problems in later life. Freud said they were the cause of *all* neuroses! It is an interesting theory. There is no evidence that it is true.

I said phobias in childhood were not related to mental illness. I want to hedge a bit on that.

Recently, psychiatrists have become aware that children sometimes have depressions like adults: clinical depressions — meaning a mood state extended over weeks or months; disabling; and having the full complement of depressive symptoms seen in adults: insomnia, sadness, guilt, suicidal thoughts. Parents are used to glum, sulky children. If every glum, sulky child were taken to a psychiatrist, most parents would be bankrupt. Well, *some* glum, sulky kids are depressed, truly depressed, like grandmother was when she stayed in bed all day, refused to eat, and finally had to be taken to the hospital.

How do you know whether *your* glum child is depressed in this sense? If the glumness hangs on, consider it a possibility.

There are good treatments for depression (Chapter 9). Phobias are sometimes a symptom of depression.

Phobias in childhood

School phobia

School phobia is an exaggerated fear of attending school. It occurs in children of all ages, peaking around 11 or 12. Both sexes are affected, girls perhaps somewhat more than boys. It is encountered in children of every social class and the full range of academic abilities.

The first sign of a school phobia is often a physical complaint: stomach ache, sore throat, headache. The child complains of being too ill to attend school and when the mother finally gives in and says he or she can stay home, the complaint almost magically disappears. If the mother does not give in, the illness may persist until the school nurse sends the child home. The symptoms go away only to return on the next morning of a school day.

After this continues for a time and the pediatrician finds nothing wrong, the diagnosis becomes 'school phobia' and the school counsellor takes over. What might the counsellor learn?

School phobia can almost always be traced to one of two causes: a fear connected with school or a fear connected with home. School can be frightening in many ways. The teacher may tend to be frightening. Gym may be frightening, particularly undressing in front of others. Reciting in class may be frightening, often associated with a fear of fainting. Children may be teased about their looks or dress and fear going to school because of this (mothers who dress their children oddly increase this risk). Children may develop toilet phobias and want to stay home where there is privacy. Finally, any change from the status quo, particularly for the timid child, can lead to phobia, the most common examples being changes to new schools with new faces and challenges.

Much more frequently, however, what seems to be school phobia turns out to be a fear of leaving home. There has been a death in the family, or separation or divorce, or sometimes a new baby. Or maybe the parents argue a lot and there is finan-

Phobia: the facts

cial trouble. Maybe none of these things has happened but the child is over-dependent on the mother. In this case, school is not what is feared but separation from the mother. This sometimes shows up when the child, forced to go to school, insists on repeatedly calling the mother on the telephone.

Sometimes it is hard to distinguish school fear from separation fear, and here is how one counsellor handled the problem:

Billy, 11 years old, had been absent from school for seven weeks. After being absent for three weeks because of illness, he had developed a phobia towards school. He became anxious, unable to eat, and complained of chest pains. Neither punishment nor bribes led to a return to school. He said he was unaware of reasons for his fear of school.

Billy was the youngest of two children, and was described as submissive, tense, perfectionistic, and very attached to his family. He had many physical illnesses and his mother was overprotective. Increased school competition and many absences led to poor grades. He was teased by peers because of academic difficulties and his weak and sickly appearance.

Over many hours, the counsellor conducted what is described as a 'behavioral assessment procedure'. This meant asking Billy to visualize and describe a typical school day, noting indications of anxiety. These included flushing of the skin, body movements, muscular tension, vocal tremors, and tears. Mathematics and literature both were sources of anxiety. Being called on, unable to answer questions, and being teased evoked intense anxiety. Billy showed no signs of anxiety when asked to visualize getting up in the morning, having breakfast, and preparing to leave home. The counsellor concluded that the school phobia was indeed a phobia of schools.

Of the next 100 children who move up to a new school year, two or three will develop phobias related to school sufficient to concern the parents and teachers and, ultimately, interfere with the child's school performance. What is there about these

Phobias in childhood

two or three children which distinguishes them from the unaffected majority?

In studies, one encounters numerous examples of children, mothers, and families 'typically' associated with school phobia. It is said, for example, that children with school phobia commonly overvalue themselves and their achievements. When their own estimate of themselves is threatened, they become anxious and withdraw from competition, often seeking closer contact with the mother. Their estimate of themselves is not securely held, and consequently they are sensitive to threat, such as change to another class or new school, return to school after an illness, a minor episode which leads to embarrassment, and actual or fantasied academic or social failure.

The 'typical' mother develops unusually close dependence on and from their children as a compensation for their unsatisfactory marital or other relationships. The mothers themselves often have a history of an unhappy relationship with their own parents. Such mothers are reluctant to leave their children in school and will tell the teacher, in the child's presence, 'You won't be able to get him to leave me'.

In the 'typical' family one sees the over-dependent child and over-protective mother who is over-dependent herself on a husband who is impatient with both of them. Usually, school phobia is reported in small families, but in one study the opposite was true: families with the phobic child lived with the grandparents or telephoned them daily.

Because one version of 'typical' varies so often from others, I have put the word in sceptical quotes. School phobia develops for reasons which nobody understands and words like 'over-dependency' help little in explaining why a small number of children develop the problem and most do not.

School phobia must be distinguished from truancy. Many children do not go to school simply because they prefer not to. They prefer doing something else, video games today being high on the list. Truant children are often delinquent in other

Phobia: the facts

ways, getting into trouble with authorities, running away from home, and generally being hard to manage. Most inmates in penitentiaries were truant as children. Very few school phobics end up in penitentiaries.

On extremely rare occasions, the child who avoids school (and usually most other social activities) becomes schizophrenic. They are loners first and later frankly psychotic. Children with school phobia usually have friends, although sometimes their school phobia will expand to include a phobia toward parties and visits to relatives or other children's homes.

Children can become depressed like adults. When school refusal is accompanied by irritability, crying, insomnia, and concern about death, depression is probably more likely than simple school phobia.

How do children with school phobia get along later in life? Most studies show that they do about as well as other children, perhaps being somewhat on the sensitive, cautious side, but suffering from no diagnosable mental illness. This benign prospect applies more to cases where the phobia developed early in life rather than in adolescence. When school phobia first develops in adolescence, the likelihood is increased that school phobia is not the only explanation, but that schizophrenia, depression, or criminality will ultimately be the outcome.

Most experts agree that the best treatment for school phobia consists of getting the child back in school as early as possible, by whatever means. Determining the source of the phobia (e.g. teasing) will of course be useful if it can be removed. Even when the source is not clear, however, a firm mother and equally firm and understanding teacher can gradually cajole a child into attending class, sometimes by having him sit for a few days in the library or Principal's office, attending class briefly and then for increasingly longer periods, and sometimes with the mother attending class with him for a few days. Everyone should evince confidence that the child will naturally return to school and the only question is when. Failure to return to school in a reasonable time leads to a

Phobias in childhood

worsening of the phobia, as the child falls steadily behind in his school work, builds up resentment, and feels even more different from his classmates than in the beginning.

There is evidence that antidepressant drugs are helpful for school phobia, as will be discussed in Chapter 9.

7

Phobias in other mental disorders

Diagnosis precedes treatment.
                                        Russell John Howard

This book has described four phobic disorders: simple phobia, social phobia, agoraphobia, and school phobia. Each is dominated by one or more phobias; each has a characteristic age of onset, 'natural history', and associated features.

Phobias, however, may occur as symptoms of other psychiatric conditions — especially panic disorder, obsessive-compulsive disorder, and depressive disorders. As the names suggest, panic attacks dominate the first, obsessions and compulsions the second, and depressed mood the third. Phobias, if they occur, are overshadowed in importance by the dominant symptoms of these other disorders.

Terms used in psychiatry are confusing. Different terms are applied to the same phenomena, and they change from time to time and differ from country to country. Psychiatrists have trouble keeping up with them. The terms in this book come mainly from the current official US nomenclature, with some exceptions, such as school phobia (newly tagged 'separation anxiety disorder' by the never-satisfied Americans). Synonyms for the various conditions can be found in the Notes at the back of the book.

Panic disorder, obsessive–compulsive disorder, and depressive disorders are now discussed. Phobias occurring in any of these three disorders have an entirely different significance — for both outcome and treatment — than phobias occurring in a predominantly phobic disorder. The reader's understanding of phobia is not complete without knowing the main features of these three conditions.

Phobias in other mental disorders

Panic disorder

Panic disorder is a chronic condition manifested by attacks of acute anxiety usually occurring in the absence of a fear-provoking situation. It is one of the most common psychiatric disorders, affecting about 5 per cent of the adult population. Women more often have the condition than men.

Attacks usually begin suddenly, sometimes in a public place, sometimes at home, perhaps awakening the patient from sleep. There is a sense of foreboding, fear, and apprehension; a sense that one has become seriously ill; a feeling that one's life may be threatened by some illness. Some patients have the disturbing sense that their body has changed or become distorted, the 'depersonalization' described in Chapter 5.

Laboured breathing, smothering feelings, palpitations of the heart, blurred vision, tremulousness, and weakness usually accompany the apprehension and foreboding. If a patient is examined during a panic attack, signs of distress will be present: rapid heartbeat, sweating, shallow breathing, tremor, hyperactive reflexes, and dilated pupils. An electrocardiogram taken during the episode will be essentially normal.

Panic attacks vary in frequency. Some people experience them daily; others have them only once or twice a year. Other symptoms may occur between attacks, such as fatigue and nervousness.

Heart and respiratory symptoms are the most frequent complaints reported to physicians: 'I have heart spells', 'I think I'll suffocate', or 'I can hardly breathe.' The patient usually considers his illness medical in nature and the physician has trouble convincing him otherwise. Emergency rooms in hospitals are frequently visited by people with panic disorder who believe they are having a heart attack.

Symptoms may become attached to specific situations which the patient then tries to avoid. For example, he may

Phobia: the facts

choose aisle seats in theatres, close to exits, so that if an attack occurs, his escape will not be blocked. Or he may avoid social situations in which an attack would be embarrassing. But phobias play a minor role in panic disorder; the attacks usually occur for no apparent reason.

Panic disorder almost always begins in the teens or twenties. Some patients may remember the exact time and circumstance of the first attack. Some remember being awakened at night by their first panic attack. Others remember having their first attack at times of stress, such as making a speech in class. Thus, the disorder may begin suddenly with all the symptoms of on anxiety attack, but it can also begin insidiously with feelings of tenseness, nervousness, fatigue, or dizziness for years before the first full-fledged anxiety attack is experienced.

The patient's initial medical contact is not always helpful. If he comes to a physician complaining of heart and respiratory symptoms, fearful of heart disease, a physician unaquainted with panic disorder may support the patient's fears by referring him to a specialist and admonishing him to avoid exercise.

Sometimes patients complain of other problems than heart and respiratory symptoms. They may have symptoms of 'irritable colon', such as abdominal cramping, diarrhoea, constipation, nausea, belching, flatus (passing gas), and trouble swallowing which may prompt them to consult a gastroenterologist.

Panic disorder can occasionally be severe but usually the course is fairly mild. Symptoms wax and wane in an irregular pattern which may or may not be associated with events and circumstances interpreted by the patient as stressful. Despite their symptoms, most people with panic disorder live productively without social impairment. Over a 20-year period, more than half of patients recover or are much improved; about one in five continue to have moderate to severe disability.

Panic disorder rarely leads to hospitalization. It is not

Phobias in other mental disorders

associated with suicide. Probably more cardiologists and gastroenterologists see people with panic disorder than psychiatrists. When psychiatrists see them, they are usually suffering from a depression or alcoholism.

People with panic disorder do not differ from the general population in level of educational or socioeconomic status. There is no evidence that any specific type of childhood experience such as bereavement predisposes to the disorder.

Recently panic disorder has been associated with a benign minor heart condition called mitral valve prolapse. The reason for the association is not clear; possibly patients with this condition sense some cardiac irregularity and this may provoke a panic attack.

In many patients simple reassurance is the only treatment needed. Tranquillizers such as Librium and Valium may be helpful, but there is evidence that a class of antidepressant drugs called tricyclics are even more useful in preventing panic attacks (Chapter 9).

Obsessive-compulsive disorders

Obsessions are persistent distressing unwanted thoughts or impulses that are distressing, senseless, and irresistible. *Compulsions* are acts resulting from obsessions. Obsessive-compulsive disorder is a chronic or recurring illness dominated by obsessions and compulsions. The condition is far less common than panic disorder, occurring in less than 1 per cent of the adult population. Both sexes are affected equally.

Obsessionals are even more secretive about their symptoms than are phobics. If they seek professional help it is usually because of depression. Sometimes psychiatrists will see them for years without suspecting they have obsessions.

Obsessive-compulsive disorder almost always begins in adolescence or the early twenties. The symptoms are extremely diverse. A common complaint is the persistent and irrational

Phobia: the facts

fear of injuring oneself or another person, often a child or close relative. Fearful of losing control, the patient may develop avoidances or rituals which lead in turn to social incapacity. Perhaps he will refuse to leave the house, or will avoid sharp objects, or wash repeatedly to destroy 'germs'.

The obsessional may recognize that the thought is illogical, but not always. Sometimes the ideas are not, strictly speaking, illogical (germs do produce disease), and sometimes even when obviously absurd to others, the ideas are not seen this way by the obsessional. What distinguishes an obsession from a delusion is not so much 'insight' (recognizing the idea's absurdity) as the person's struggle against the obsessional experience itself. He actively strives to resist the obsession, to free himself from the thought, but cannot and feels increasingly uncomfortable until the idea temporarily 'runs its course' or the obsessional act has been completed.

The illness may assume one or more of the following forms:

Obsessional ideas. Thoughts which repetitively intrude into consciousness (words, phrases, rhymes), interfering with the normal train of thought and causing distress to the person. Often the thoughts are obscene, blasphemous, or nonsensical.

Obsessional images. Vividly imagined scenes, often of a violent, sexual, or disgusting nature (images of a child being killed, cars colliding, excrement, parents having sexual intercourse) that repeatedly come to mind.

Obsessional convictions. Notions that are often based on the magical formula of thought-equals-act ('Thinking ill of my son will cause him to die'). Unlike delusions, obsessional beliefs are characterized by ambivalence: the person believes and simultaneously does not believe. As a famous psychiatrist, Karl Jaspers, expressed it, there is a 'constant going on between a consciousness of validity and non-validity. Both push this way and that, but neither can gain the upper hand.'

Phobias in other mental disorders

Obsessional rumination. Prolonged, inconclusive thinking about a subject to the exclusion of other interests. The subject is often religious or metaphysical — why and wherefore questions which are as unanswerable as they are impossible to resolve. Indecisiveness in ordinary matters is very common ('Which necktie should I wear?'). Doubt may lead to extremes in caution both irksome and irresistible ('Did I turn off the gas?' 'Lock the door?' 'Write the correct address'?). The patient checks and rechecks, stopping only when exhausted or upon checking a predetermined 'magical' number of times. Several studies suggest that obsessional doubts — *manie du doute* — may well be the most prominent feature in obsessional-compulsive disorder.

As with other obsessions, ruminations are resisted. The person tries to turn his attention elsewhere, but cannot; often the more he tries, the more intrusive and distressing the thoughts become.

Obsessional impulses. Typically relating to self-injury (leaping from a window); injury to others (smothering an infant); or embarrassing behaviour (shouting obscenities in church).

Obsessional fears. Often of dirt, germs, contamination; of potential weapons (razors, scissors); of being in specific situations or performing particular acts.

Obsessional rituals (compulsions). Repetitive, stereotyped acts of counting, touching, arranging objects, moving in specific ways, washing, tasting, looking. Compulsions are inseparable from the obsessions from which they arise. A compulsion is an obsession expressed as action.

Counting rituals are especially common. The person feels compelled to count letters or words or the squares in a tiled floor, or to perform arithmetical operations. Certain numbers, or their multiples, may have special significance (he 'must' lay down his pencil three times or step on every fifth

Phobia: the facts

crack in the pavement). Other rituals concern the performance of excretory functions and other everyday acts such as preparing to go to bed. Also common are rituals involving extremes of cleanliness (handwashing compulsions, relentless emptying of ashtrays) and complicated routines assuring orderliness and punctuality. Women apparently have a higher incidence of contamination phobia and of compulsive cleaning behaviour than do men.

A newspaper story tells of a 49-year-old Australian housewife who is

tortured by one of the world's worst cases of fear of germs — the same bizarre phobia that plagued the late billionaire Howard Hughes.

In a fanatic pursuit of cleanliness she uses up more than 225 bars of soap on herself every month, wears rubber gloves even to switch on a light — and makes her husband sleep alone so she won't be contaminated by him.

Every month, Mrs. X goes through 400 pairs of surgical gloves, 4,000 plastic bags — which she wears in multiple numbers over the gloves — and 360 rolls of paper towels.

She goes through dozens of boxes of laundry detergent every month because she washes her clothes six or seven times before wearing them. 'And I can't bear to walk on the floors outside my bedroom. I spread newspapers ahead of me as I walk through the house. But I can't stand leaving them lying on the floor — so I leave a room by walking backward and picking up the papers in my gloved hands.

I'm terrified of encountering dirt. Whenever I feel particularly uneasy about dirt I wash my hands. Once I start I can't stop until I've used up the entire bar. That usually takes about 90 minutes.'

The woman's account may seem exaggerated, but probably isn't. Obsessive–compulsive disorder may be utterly devastating.

Four kinds of rituals occur most frequently: counting, checking, cleaning, and avoidance rituals. Avoidance rituals are similar to those seen in phobic disorders. An example can be seen in a patient who avoided anything coloured brown.

Phobias in other mental disorders

inability to approach brown objects greatly limited his activities.

Other rituals which occur less often consist of 'slowing', 'striving for completeness', and 'extreme meticulousness'. With slowing, such simple tasks as buttoning a shirt or tying a shoelace might take up to 15 minutes. Striving for completeness may be seen in dressing also. Asked why he spends so much time with a single button, the patient might reply that he was trying to prove to himself that he had 'buttoned the button properly'.

A common form of pathological meticulousness is a concern that objects be arranged in a special way. Pencils, for example, may have to be arranged so that the points are directed away from the patient. Students may spend so much time in arranging pencils, pens, erasers, etc. that they cannot do their work.

Rituals, ridiculous as they may seem to the patient, are accompanied by a profound dread and apprehension that assure their performance, since they alone give relief. 'I'll explode if I don't do it', a patient may say. Occasionally a patient believes that failure to perform a given ritual will result in harm to himself or others, but often the ritual is as inexplicable to the patient as it is to the observer.

Obsessional symptoms often are accompanied by depressed mood. This may lead to an erroneous diagnosis of depression, since the mood element at the time of examination may overshadow the obsessional content.

Obsessional symptoms rarely occur singly. As with most psychiatric illnesses, obsessive–compulsive disorder presents a cluster of symptoms which, individually, are variable and inconstant over time but as a group maintain characteristics unique to the illness. Thus a patient may now have one set of obsessional impulses and rituals, later another set, but the symptoms remain predominantly obsessional.

The illness may begin gradually or suddenly, usually for no apparent reason.

Phobia: the facts

Obsessionals with mild symptoms requiring only outpatient therapy have a rather good prognosis; as many as 80 per cent are recovered or improved five years after diagnosis. Hospitalized cases do less well. One-third or fewer are improved several years after discharge, and between 5 and 10 per cent experience progressive social incapacity.

Despite the frequency with which suicide may figure in obsessional thinking, obsessionals rarely commit suicide. Obsessional patients sometimes fear they will injure someone by an impulsive act. They fear they will lose control and embarrass themselves in some manner. They worry about being addicted to drugs prescribed by their physician. These fears are generally unwarranted. There is no evidence that obsessive-compulsive disorder predisposes to homicide, criminal behaviour, alcoholism, or drug addiction.

Finally, obsessionals may fear they will 'lose' their minds, become totally disabled, need chronic hospitalization. None of these events is common.

Obsessive-compulsive disorder must be distinguished from 'obsessional personality'. No investigator has followed a group of clearly defined obsessional personalities over enough time to determine their fate; hence the label has no predictive value and is not in this sense a diagnosis. The individual with an obsessional personality is punctual, orderly, scrupulous, meticulous, and dependable. He is also rigid, stubborn, pedantic, and something of a bore. He has trouble making up his mind, but, once made up, is single-minded and obstinate. Many individuals with obsessive-compulsive disorder have obsessional personalities antedating the illness.

It is not known whether obsessional personalities more often develop phobias than other people, but people with obsessive-compulsive disorder clearly do. The avoidances, however, usually have a ritualistic, compulsive, almost *ornate* quality that obviously goes far beyond simple avoidance.

Phobic disorders are easier to treat than obsessive-compulsive disorder. A class of antidepressant drug called

Phobias in other mental disorders

MAO inhibitors have been reported useful for obsessions and compulsions, but not very convincingly. For a time, lobotomy seemed to have the most dramatic therapeutic results, but this treatment has fallen into almost total disuse in recent years — many think wisely, because of side effects.

Depressive disorders

Feelings of sadness and grief are universal and why shouldn't they be? As the French philosopher, Amiel, expressed it:

Melancholy is at the bottom of everything, just as at the end of all rivers is the sea . . . Can it be otherwise in a world where nothing lasts, where all we have loved or shall love must die? . . . The gloom of an eternal mourning enwraps every serious and thoughtful soul, as night enwraps the universe.

Sadness and grief must be distinguished from clinical depressions, meaning depressions which lead to disability. Night may 'enwrap the universe', but clinical depressions are fairly uncommon. Perhaps 3 or 4 per cent of people at some point in their life have one.

This section refers to depressive 'disorders' in the plural. This is because every classification system divides depression into two or more types. The world consists of lumpers and splitters. The lumpers are usually content with two depressions but the splitters may have as many as a dozen. (The American classification now includes six.) There are many names for depression: manic–depressive illness, endogenous depression, melancholia, psychotic depression, affective disorder, bipolar illness, unipolar depression, secondary depression, reactive depression, and on and on (see Notes). In truth, clinical depression may be a single phenomenon with varied manifestations. The cause of clinical depression(s) is not known and as long as this is true the decision to lump or split is a matter of taste more than science.

I have opted for two depressions — primary and secondary

Phobia: the facts

— defined as follows: primary depression occurs in people who are previously well; secondary depression occurs in people who already have a psychiatric illness or a serious medical illness. The symptoms may be identical. The chief justification for dividing depression into primary and secondary types is that the primary type is associated with a high risk of suicide and the secondary type is not.

Except for mentioning conditions in which secondary depressions commonly occur, the distinction will not be referred to again. These conditions are schizophrenia, diseases of the brain, alcoholism, drug dependence, obsessive-compulsive disorder, agoraphobia, endocrine disorders, and chronic debilitating medical illnesses such as diabetes and heart disease. Interestingly, terminal illnesses, even when much pain is involved, are usually *not* associated with clinical depressions. Cancer victims very seldom commit suicide.

The symptoms of depression are *psychological* and *physical*. The psychological symptoms include feelings of sadness, despair, guilt, hopelessness, and self-loathing. There is trouble concentrating. The person becomes preoccupied with death. He may contemplate suicide and even make preparations for suicide. He may develop obsessions identical to those described in the preceding section on obsessive-compulsive disorder. He may develop phobias.

The phobias are those usually associated with agoraphobia: a fear of leaving the house and a fear of crowds and confining places. Depressed patients often develop symptoms of agoraphobia, and agoraphobics are often depressed, leading some authorities to believe they are the same illness. Supporting this idea are studies showing a high prevalence of depression in the families of agoraphobics. Against the idea is the low suicide rate in agoraphobia; the fact that agoraphobia tends to be chronic and depression to be episodic; and the fact that agoraphobia is a disorder overwhelmingly dominated by phobias, whereas the phobias in depressive disorders tend to be

Phobias in other mental disorders

fleeting and overshadowed by the depressed mood and feelings of guilt and worthlessness.

In any case, more than half of depressed patients have one or more phobia-like symptoms.

One symptom in depressive disorders stands out above the rest: loss of interest. Interest is the stuff that keeps us going, as expressed in this bit of doggerel by Richard Condon:

> Interest is the key to life
> Interest is the clue
> Interest is the drum and fife
> And any God will do.

In depression, there is no key, clue, drum, or fife. Nothing is enjoyed any longer: food, sex, hobbies, work, companionship, even a nice day. Without enjoyment, a person sees no reason to go on. He feels a profound sense of emptiness. Suicide becomes perfectly comprehensible. 'There is no future' is perhaps the most common observation made by depressed patients.

Depression may also include psychotic symptoms, referring to hallucinations and delusions. The patient hears imaginary voices, often accusing him of something. He may see visions of a dead relative or the Devil. He may have delusions.

A delusion is a fixed false idea. No amount of persuasion or evidence shakes the person in his delusional belief and the belief is regarded as false by everyone, or almost everyone. (In the case of religious beliefs it is sometimes hard to tell what is delusional from what isn't because it is hard to know what is false. However, those who share the religious ideas of the patient usually recognize a pathological form of the belief.)

Delusions in depression tend to fall in certain 'understandable' categories: delusions of poverty, delusions of bad health, and delusions of worthlessness. These are understandable in that most normal people are concerned about money, health, and what others think of them. But a depressive delusion goes far beyond normal concern. The depressed person is

Phobia: the facts

convinced he has cancer despite all evidence to the contrary; he believes he is impoverished despite a healthy bank account; he believes he has sinned against others when in fact he has been a paragon of virtue.

Old people with psychotic delusions have been known to return a few shillings or dollars to the government, convinced they cheated on their taxes many years before.

People with depressions commonly fear they are losing their mind and, in the case of psychotic depressions, they indeed have. However, their 'mind' returns when the depression ends.

And the depression always ends — or almost always. Depression is an episodic illness, as will be mentioned again a little later.

But to complete the list of symptoms first: depression, like anxiety, always has a physical dimension. The patient cries. He looks older and sometimes smaller. He walks and talks slowly or sometimes the reverse: he becomes agitated, pacing the floor, wringing his hands, and carrying on about what a bad person he is.

A sleep disorder is almost always present — either insomnia or oversleeping. (If a person claims to have a depression and sleeps well, he is probably taking sleeping pills or is not depressed.) He loses his appetite, he loses weight. He is constipated and sometimes has trouble urinating; all of his physiological functions seem to slow down. He loses interest in sex and has trouble performing on demand.

And he has medical complaints, the reason many depressed patients finally see a physician. Little pains become big pains. There is a loss of energy (the physician automatically thinks of thyroid disease). Headaches and indigestion bother people never bothered by them before.

When the depression goes away, the medical problems disappear.

As mentioned, depressions do generally go away. At the time they are depressed, people cannot appreciate this. One of

Phobias in other mental disorders

the hardest things a psychiatrist tries to do is to convince the suicidal depressed patient that he will eventually stop being depressed and that the reason for suicide will go away.

Some depressions are said to be chronic, and this may be true, but they are very much in the minority.

Depressions can occur at any time in life (although whether children experience depression in the same sense as adults do is a subject of controversy).

Most suicides in Western countries are committed by people with 'primary' depressions. One reason to believe our treatment for depression is still less than perfect is the fact that the suicide rate in Western countries has not especially declined, despite many new pills on the market for depression.

About grief reactions: people who have lost a loved one by death or separation may have all of the symptoms mentioned above except they rarely fear they are losing their mind and rarely commit suicide. They also rarely see psychiatrists or other physicians for their depressive symptoms, probably because their depression seems justified. Primary depressions come out of the blue. Often the victim literally does not know what is happening. The mysteriousness of the condition causes anxiety, and many if not most depressions have elements of anxiety, even to the extent of resembling a panic disorder.

Perhaps a quarter of the people who have depressive disorders also experience episodes of mania, explaining the existence of the term 'manic-depressive disease'. Mania is manifested by hyperactivity, rapid thoughts, euphoria, and often delusions of the grandiose type (the person thinks he is Christ or Napoleon). For nearly 100 years it has been assumed that people who experience both depressions and mania have one illness — manic-depressive disease — with a single cause, whatever the cause may be. After a hundred years it is still not known whether this is true.

The treatment for depression is pretty good (despite a non-declining suicide rate). Antidepressant drugs work in a

Phobia: the facts

majority of patients. When they fail, electroconvulsive therapy often produces dramatic results. A good deal of scepticism and even enmity have been directed towards the latter treatment but its effectiveness in depression has been documented to a degree unequalled in psychiatry.

8

The cause of phobias

> I have come to the conclusion that my subjective account of my own motivation is largely mythical on almost all occasions. I don't know why I do things.
>
> J.B.S. Haldane

> The wilderness masters the colonist.
>
> Frederick J. Turner

Phobias do not occur at random. Any explanation for them must answer these questions:

1. Why do some people develop phobias and most do not?
2. Given so many things to fear, why does a person become phobic about a *particular* thing?
3. Why are some phobias especially common?
4. Why do certain phobias occur at certain ages?
5. Why do women develop phobias more than men?

There are many theories about the cause of phobias, but none satisfactorily deals with these questions. <u>The cause of phobias remains unknown</u>.

The theories are interesting, however, and some may even be true, in part. Theories are often viewed as mutually exclusive, but this is a mistake. Conceivably, if some genius could identify the true parts and synthesize them into a theory which could be tested scientifically, the cause or causes of phobia might become clear. No synthesizer has appeared, and we are left with theories bumping shoulders but rarely speaking to each other.

Two theories have dominated the others in this century. One originated with Sigmund Freud. John Watson proposed the

Phobia: the facts

other (although Pavlov deserves the real credit). As the years passed, both theories were modified somewhat by their originators and followers. They are presented here in more or less their original form on the grounds that the modifications did not modify much and the conviction that the reader would prefer simplicity to stupor.

Both theories were named after little boys, which may help keep them straight.

Little Hans was a five-year-old when Freud studied his case. Hans had a horse phobia. After seeing him once and talking with the father, Freud traced the phobia to this sequence of events:

1. Hans lusted for his mother.
2. He was jealous of his father, a rival with a larger penis.
3. He could not express his jealousy because of fear, specifically fear that he would be punished by the father for lusting for the mother. The punishment would take the form of castration (a punishment he assumed was common, based on his observation that little girls lacked penises, presumably because they, too, had lusted for a parent).
4. The fear was so great he actively 'forgot' (*repressed*) the whole business. For further protection, his Unconscious Mind performed the following manoeuvers:
5. He *projected* his hostility on to the father: the father became the aggressor. Then he *displaced* his aggressive feelings from the father to something safer, more neutral: horses.

Horses became the objects to be feared and avoided. Horses were a *symbol* for the father, but Hans did not know this. He certainly did not connect his fear of horses with *sex* — incestuous sex, at that. It took Freud, ranked by his admirers with Copernicus and Einstein as one of the three great geniuses in history, to make *that* connection.

The cause of phobias

In the little Hans story are all the ingredients of the psychoanalytical theory of phobias: sexual feelings toward the parent of the opposite sex combined with jealousy of the parent of the same sex (the Oedipal complex); anxiety arising from the anticipation of castration; formation of 'unconscious defences' against the anxiety, namely, repression, projection, displacement, and symbolization.

Little Hans could much more easily avoid horses than what they symbolized: his father. Of course he paid for these unconscious manoeuvers by developing a 'neurotic' symptom: a phobia.

That is how Freud explained not only Hans's horse phobia but *all* phobias. Copernicus and Einstein were no match for this brand of audacity — if that is the word for it.

How did Freud explain phobias occurring later in life? Do men fear castration forever? What about women? Do *they* fear castration?

Freud answered yes to both. Castration fears are unconscious; the person is not aware he has them. In fact, he invests considerable psychic energy to avoid being aware he has them. Simmering in the unconscious mind, castration anxiety can erupt at any time in the form of a phobia when stress forces a person to 'regress' to 'an infantile level of functioning' (regression being a way to escape the anxieties of adult life).

Of course, if Freud or another psychoanalyst could help a phobic patient bring his castration fears into conscious awareness, the fears would be recognized as absurd and go away. When an adult first realizes he has a secret fear of being castrated, a tremendous emotional storm called an 'abreaction' may occur, but it passes quickly and the patient is cured. 'Insight' alone cures him. This was how psychoanalysts proposed to cure phobias, at least in the early days of psychoanalysis. Later, insight was downplayed as analysts realized that phobias were not being cured at the expected rate.

As for women, who, so to speak, were born castrated, penis

Phobia: the facts

envy becomes the unconscious counterpart of castration anxiety.

How this works is illustrated in two case studies reported by psychoanalysts:

Case one:
A 12-year-old boy develops a phobia of heights. As a child, he cried in the presence of a friend of his father who straightaway held him over a cask of water and threatened to drop him in unless he stopped crying. Later, a teacher held him upside down over a high wall and threatened to drop him. Thereafter he avoided heights of any kind.

The analyst rejected the idea that stressful experiences alone produce a fear of height, and 'the notion of a congenital predisposition to develop fear reactions . . . is an impossible assumption'. Instead he presented a brief analysis of the case in which he assumed that the child had hostile thoughts bred by jealousy of his father and that when the teacher — whom he identified with his father — held him over the wall, his thoughts were: 'It's come at last — the all-knowing father has discovered my sinful thoughts and is going to do to me what I wished to do to him.'

His fear of heights was intelligible, the analyst believed, only if it was remembered that in the child's unconscious mind the avenging father was always with him; and the fear that he himself might fall from a height was a repressed desire to do wrong — to 'fall' (morally).

Case two:
A woman in the midst of divorcing her husband developed a crowd phobia and would not leave the house. Her fear was traced by her analyst to a 'deep hostility originally directed toward her mother, then deflected on to herself'. 'Everybody look!' her anxiety seemed to proclaim. 'My mother let me come into the world in this helpless condition, without a penis'. Rather than let the world see her without a penis she stayed home.

Freudians later decided that castration anxiety and penis envy were not the sole source of phobias. Other unconscious fears also expressed themselves in a symbolic manner as a phobia. School phobia might develop from the unconscious fear of killing one's mother. A desire to exhibit oneself may lead to a fear of crowds. Open streets may assume various

The cause of phobias

symbolic meanings: an opportunity for sexual adventure or the idea that some other person (usually a parent or sibling) may die while one is away from home. Street traffic may symbolize parental intercourse. Women may develop a fear of shopping because shopping symbolizes taking the mother's place beside the father.

Sometimes the defence against fear is to do what is feared — to the limit! The person afraid of heights becomes a sky-jumper; the person with a cat phobia becomes a lion tamer. The person now has what is called a 'counterphobia'. Theoretically, fears go away when confronted, but the Freudians say no, the *real* fear is not of heights or cats but of castration, etc., and this fear is not being confronted.

As possible sources of anxiety expanded, the possible symbolic expressions of anxiety became increasingly varied. Without abandoning Freud's original idea that phobias represented a 'defence against anxiety', latter-day Freudians speculate that even adult non-sexual sources of anxiety may produce phobias, given an essential point: the person must remain unaware of the source of his anxiety. 'The fears we know are of not knowing', W.H. Auden said, and a Freudian theme that seems to make sense is that people would rather fear something specific and definite than experience 'free-floating anxiety', where the person has no idea of the source of his fear. From this point of view the mind can be viewed as a factory forever converting the raw material of free-floating anxiety into concrete fears which can be dealt with, even if the fears are absurd, such as fear of open streets, tomcats, or men in pink shirts.

Freudian theory does not lend itself to scientific testing. In science one can never prove that something is true; one tries to prove that something is untrue, hoping that after the untruths have been eliminated, what remains may possibly be true.

It is impossible to disprove the theory that phobias arise from castration anxiety. Freudians do not attempt to prove it. They listen to patients and select from the patients' remarks

Phobia: the facts

those which fit their theory. William James, the famous American psychologist, pointed out that you can toss a bag of marbles on the floor and by selectively ignoring certain marbles find any pattern you wish. Not only Freudians do this, of course; as we will see with the Behaviourists, the practice is widespread.

Little Albert was 11 months old, a phenomenally unperturbable infant. You could put him in a room with rats, rabbits, dogs, and other menacing creatures and he would not blink an eye. A loud noise would make him jump and cry, but that was all. Fearless.

It was 1920 and Dr. John B. Watson was interested in fear. Freud's theories were popular at the time but Watson was a dedicated Pavlovian and did not like Freud's theories. He was sure he could produce fear even in a rock like little Albert, and do it without introducing seductive mothers or scissors-toting fathers.

He succeeded.

Dr. Watson took advantage of Albert's startle response to loud noises and the principles of conditioning discovered by I.P. Pavlov. He showed Albert a rat and then banged an iron bar behind his head. Albert jumped and crawled away from the rat. After several exposures to the rat–bang combination, the bang wasn't needed. Just seeing a rat made Albert cry and race for the opposite side of the crib. Albert had a rat phobia, and John B. Watson was responsible!

Then something unexpected happened. Albert showed the same fear response to almost anything furry: a rabbit; fur coat, even a mask of Santa with a beard. Santa had become a *symbol* for rats. The good behaviourist Watson rejected words like 'symbol' but for Freudians it was symbolism, pure and simple, with little Albert 'displacing' his new-found fear of rats on to other neutral objects which could then be avoided.

The rat could be avoided too, if Watson did not keep

The cause of phobias

putting it in the crib. He did keep putting it in the crib, however, without banging the bar, and after a while Albert recovered from his rat phobia. Repeated exposure to the phobic object is the road to cure, as Chapter 9 makes clear.

But, after Albert was cured, Watson pulled one of his nasty tricks. He brought in the rat and once again banged the bar. He banged it only once but it was enough to make Albert terrified of rats again.

Phobias have this peculiarity: they go away but sometimes return, often for no apparent reason. You think you have the patient cured and there he is, a year later, avoiding men in pink shirts again. It is as if, once a phobia has settled in, it can be dislodged, but the possibility of relapse persists indefinitely. A sudden loud noise — a brief panic in the supermarket queue — and the phobia is in business again. Treating phobias is notoriously frustrating.

Watson was the father of behaviourism, a branch of learning theory. Learning theory holds that fears are conditioned responses and thus learned, just as Pavlov's dog 'learned' to salivate when he heard a bell. The responses occur if the responder is rewarded but not if the responder suffers (Skinner's contribution to learning theory). What, then, maintains a phobia? What is the reward?

The reward is powerful indeed. Every time a person avoids a phobic situation, he escapes fear. Avoidances may be a nuisance, but clearly preferable to panic.

Learning theory is simpler than Freudian theory. Freud also believed phobias were learned, but by a tortuous, convoluted route. A conditioned response is comparatively straightforward. A loud noise or other stressful stimulus produces an 'unconditioned' response: fear. Pair the stressful stimulus repeatedly with a harmless object, person, or situation and eventually the harmless object, person, or situation produces a 'conditioned' response: fear. If, to avoid fear, the person avoids the harmless object, person, or situation, the fear becomes a phobia.

Phobia: the facts

Fear can be conditioned to literally anything. In the case of chronically anxious people, Behaviourists suspect that perceptions of their own bodies or patterns of light and shade become conditioned stimuli eliciting anxiety. Whenever *anything* happens to us, *something* is going on simultaneously, if only our heartbeat and the ambient temperature. If we are in danger and feel justified fear, our heartbeat and the ambient temperature, by their quite coincidental, simultaneous presence at the time of danger, may later produce *unjustified* fear, or phobia.

Learning theory is not only simpler than Freudian theory but also provides more grounds for therapeutic optimism. The basic tenet of learning theory is that whatever is learned can be unlearned. Habits can be broken, conditioned responses extinguished. The secret of unlearning a fear is to expose oneself repeatedly to the feared situation and eventually, if no harm comes to one, the fear will go away. Freudian analysts are less optimistic. The fears go back to childhood; they lurk deep in the Unconscious Mind and have many disguises; it may take years to root them out, if ever.

Freud himself took a pessimistic view of the treatment of phobias. He himself had a travel phobia which persisted even after his self-analysis. He agreed with the behaviourists that the ultimate cure for phobia was to confront the phobic situation and keep it up until the phobia disappeared. As for himself, he stayed off trains as much as possible.

But the behaviourist theory of phobias also has flaws. The principal one is that everybody goes through life surrounded by danger. Living is a risky business for all of us. Why don't all of us develop phobias? As the behaviourists say, the opportunities for conditioning are limitless. Our very bodies can become a conditioned stimulus eliciting fear — not to mention cinemas, elevators, and the neighbour's cat. Even fear can become a conditioned stimulus eliciting fear. The fact is, only a minority of people develop phobias — although

The cause of phobias

perhaps a larger minority than was once suspected. There is more to the story than the chapters by Freud and Watson. Their theories are fascinating, fun, and possibly, to a degree, true. But neither theory answers a *single* question posed at the beginning of this chapter. For answers we must gather facts from other sources: however dubious, however dull.

To explain odd behaviour like phobias it is useful to divide information into two categories — one dealing with biological factors, the other with life experiences. Taking them in order:

Biological factors

As discussed in Chapter 3, some fears are innate. There is no other explanation for an adult chimpanzee, seeing a snake for the first time 20 yards away in a zoo, going into a panic. People also have innate fears: fear of strangers, fear of being stared at, fear of falling into space. These fears even follow a kind of biological timetable. At six months, the infant is frightened by loud noises and sudden movements. At age three he is frightened by strangers; at five by animals. Fear of open spaces and social situations occur (if they occur) much later: in adolescence or early adulthood. There is a remarkable consistency across cultures in the timing of these 'normal' fears, strongly suggesting biological factors.

'Innate' does not necessarily mean 'inherited'. Before a person is born, he spends nine months in the womb, the most important nine months in his life. Experiences in the womb powerfully influence postnatal appearance *and* behaviour. For example, if a female monkey fetus receives male sex hormones at a crucial period of development, the female newborn not only has masculine features but, later, behaves sexually as a male.

Descartes, in the seventeenth century, observed that if a pregnant woman smells roses or is frightened by a cat, her

Phobia: the facts

unborn child will also smell roses and may have an aversion to roses or cats 'imprinted in his brain to his life's end'. Today, geneticists do not believe in the inheritance of acquired traits, but no one doubts that what happens *chemically* to the mother also happens chemically to the newborn, and that smelling as well as fear involve chemical reactions.

Intrauterine experiences may have no relevance to postnatal phobias, but, again, they might, and Descartes' point that 'innate' is not always 'genetic' is useful to remember (see Notes, Chapter 1).

Innate fears usually do not reach phobic intensity but may lay the groundwork for a phobia if reinforced by events in later life. For example, a three-year-old is afraid of strangers but gets over it by five. Then, at 12, he bungles a class recitation and thereafter has a phobia about public speaking. At five, a child is frightened by dogs but not for long; every neighbour has one and he likes his own dog too much to be afraid of them. Then, at 20, a dog bites him and he develops a dog phobia.

Innate fears may explain why some phobias are more common than others, but obviously this does not explain the whole story. People are bitten by dogs every day and most do not develop dog phobias.

In terms of arms, legs, opposing thumbs, and maybe certain fears, we are all born with similar characteristics but also with conspicuous differences. Some legs are short, some long, some straight, some bowed, etc. The same is true of temperament. Tiny babies have strikingly different temperaments, ranging from stolid to tempestuous. The differences cannot be attributed to differences in maternal care; obstetricians spot them at time of delivery. Our temperaments may be as moulded by heredity as our hairlines.

It was true of Pavlov's dogs. Some were easily conditioned and others not, and he blamed 'temperament'. Identical twins provide further evidence. In those rare instances when identical twins are separated at birth and raised apart in different

The cause of phobias

environments, they show striking temperamental similarities in adulthood. If one is shy, the other is shy; if one is anxious, the other is anxious; and if one has phobias, the other *tends* to have phobias.

Does this mean that specific phobias are inherited? Probably not. Specific phobias do not run in families. There is some evidence that agoraphobics have an increased rate of various psychiatric illnesses in their families, but the evidence is mixed and not impressive.

What is more likely, perhaps, is that people vary in their innate susceptibility to anxiety. Anxious people, for example, develop a conditioned response faster than unanxious people. To the extent that conditioning plays a role in phobias, anxious people may be phobic-prone. Even people with normally low levels of anxiety become anxious in certain situations or when depressed; they may be particularly ripe for a phobia during these times. (As noted in Chapter 7, clinical depressions are often accompanied by phobias.)

Anxiety produces an increase in adrenaline. However, adrenaline does not produce anxiety. There is one chemical which does make people anxious: lactic acid. This chemical, normally produced by the body, administered to people in a calm state, produces symptoms of anxiety but *only* in people who have had anxiety attacks in the past. Conceivably people who develop phobias have higher blood levels of lactic acid than people resistant to phobias, but this is not known and has not been studied.

The personalities of phobics have been often studied, with mixed results. Some studies find phobics to be timid, dependent, and immature, and others find them to be outgoing and independent. Therapists are inclined to view agoraphobics as exceptionally dependent people who are lifelong avoiders of difficult situations. Joseph Wolpe, the founder of the desensitization method of treating phobias, describes an agoraphobic in the following terms:

Phobia: the facts

An only child, during her childhood and adolescence she had been incredibly overprotected by her mother who insisted on standing perpetually in attendance on her. She was permitted to do almost nothing for herself, forbidden to play games lest she get hurt, and even in her final year at high school was daily escorted over the few hundred yards to and from school by her mother, who carried her school books for her.

Studies sometimes show phobics have dependent personalities and overprotective mothers, but sometimes do not. Whether certain personalities are particularly subject to phobias remains an open question.

Life experiences

Some people attribute their phobia to a single frightening event. This happens most often with simple phobias. A person is stuck in an elevator and avoids elevators; a dog bite leads to a dog phobia; a young person learning to drive dents the family car, is scolded, and refuses to drive again.

Obviously the traumatic event does not tell the whole story: many more people are stuck in elevators than develop elevator phobias. Most phobias, moreover, are not associated with a single frightening event — at least one which the victim recalls.

Perhaps, however, the victim has forgotten the traumatic event. Perhaps it happened when he was a small child and he repressed the memory. Or, even as an adult, perhaps people actively put out of their conscious mind thoughts which are too upsetting.

Some years ago, there was a theory, popularized in novels and Second World War movies, that *forgetting* a traumatic event was more important than the trauma itself in causing a neurotic disturbance such as a phobia. A soldier would see his buddy killed and repress the memory only to have it expressed symbolically as blindness, paralysis, or another form of 'hysteria', or, in some cases, severe phobias. The treatment

The cause of phobias

consisted of the army doctor bringing the patient's memory back to consciousness — sometimes with the help of hypnosis or sodium amytal. This dramatic cure, called 'catharsis' or 'abreaction', differed from the early Freudian cure based on 'insight'. The latter came from the theory that phobias were caused by anxiety resulting from unconscious childhood sexual conflicts; bringing the memory of these into conscious awareness was the heart of the treatment but a far more lengthy and complicated process was required than occurred on the movie screen — and sometimes in real life.

Cathartic cures are not as fashionable now, probably because they usually do not work. Today's version of 'shell shock' (First World War) and war neurosis (Second World War) is something called the 'post-traumatic stress disorder'. This usually refers to a veteran who has intrusive recollection of bad war experiences and avoids activities which might remind him of the bad experiences; he also has other symptoms, including insomnia and guilt. The treatment usually consists of antidepressant pills or other medication; the cathartic method is rarely used. The veteran's problem is not amnesia; the problem is remembering things too well.

Just as cathartic cures of amnesia have gone out of style, so has amnesia itself disappeared to some extent — at least amnesia associated with stress. In the early years of the twentieth century, amnesia was commonly seen by neurologists and psychiatrists, and the theory that neurotic disturbances resulted from amnesia — repression of traumatic memories — was widely held. The medical literature of that period is replete with case histories supporting this view — so many that one is almost compelled to take the theory seriously. Here is a case from the early 1920s:

A young boy would often pass a grocery store on errands, and when passing would steal a handful of peanuts from the stand in front. One day the owner saw him coming and hid behind a barrel. Just as the boy put his hand in the pile of peanuts the owner jumped

Phobia: the facts

out and grabbed him from behind. The boy screamed and fell fainting on the sidewalk.

The boy developed a phobia of being grasped from behind. In social gatherings he arranged to have his chair against the wall. It was impossible for him to enter crowded places or to attend the theatre. When walking on the street he would have to look back over his shoulder at intervals to see if he was closely followed.

This phobia continued until the age of 55, when the man returned to the town of his childhood. He met the grocer, introduced himself and during reminiscences the grocer finally told him the story of the stolen peanuts. The man remembered the episode and the phobia disappeared after a period of readjustment.

And here is a case from the 1950s:

While a 4-year-old girl was playing her pet dog knocked her little sister on the ground, causing a splinter wound in the cheek. The sister died a few days later, apparently of an infection. The mother openly accused the patient of knocking her sister down. The day of the sister's funeral the mother angrily accused the child again of causing yet other damage which in fact had been produced by the dog. Several days later the girl began to dislike dogs and soon developed a severe dog phobia which persisted until the age of 45, when the patient sought treatment as she wanted to give her daughter a puppy as a birthday present. Memories of the original events were recovered by the patient after four sessions of hypnosis. The dog phobia rapidly subsided and the patient remained well a year later.

One does not hear many stories like this these days. Why not? It is understandable how treatments change with time, but do symptoms also change? The answer is yes.

Culture has a strong influence on phobias. In the sixteenth century, people had phobias about witches. In the nineteenth century, phobias about venereal disease were common. Today, in the United States, people are phobic about heart disease; in France the liver seems to be the primary target; and in England, the digestive tract.

Some phobias, of course, seem a permanent fixture: fear of heights, closed places, and animals. Hippocrates wrote about them more than 2000 years ago.

Simple phobias — now and in the past — appear more

The cause of phobias

strongly related to the occurrence of a single traumatic event than do social phobias. The onset of agoraphobia is particularly shrouded in mystery. Agoraphobics rarely associate the onset of their phobias with a single frightening event, but vividly recall their *first* panic attack. Freud reported this first. In agoraphobia, he wrote, 'we often find a recollection of a state of panic; and what the patient actually fears is a repetition of such an attack under those special conditions in which he believes he cannot escape it.'

Agoraphobics rarely have an explanation for the first anxiety attack. 'It came out of the blue', they say. They associate a future anxiety attack with other unpleasant possibilities — a fainting spell, a sudden illness, something else embarrassing — but their real reason for staying in the house and avoiding open spaces and public places is to avoid another attack. What makes the prospect of an anxiety attack particularly terrifying is the unexplained nature of the first attack: if it happened for no obvious reason once, it could happen again for no obvious reason.

Studies of agoraphobia find that the first anxiety attack often occurs against a background of worry and unhappiness: job dissatisfaction, a domestic crisis, a serious medical illness, a death in the family. The relationship of non-specific stress with the onset of agoraphobia led one investigator to call agoraphobia the 'calamity syndrome'. Nevertheless, agoraphobics do not associate the first anxiety attack with a *specific* stress that would justify panic. Isaac Marks quotes a patient as saying:

I was standing at the bus stop wondering what to cook for dinner when suddenly I felt panicky and sweaty, my knees felt like jelly, and I had to hold on to the lamppost and I was afraid I would die. I got on the bus and was terribly nervous, but just managed to totter home. Since then I haven't liked to go out on the street and have never been on a bus again.

Marks comments:

Phobia: the facts

In such cases the primary event is anxiety which arose within the patient for no accountable reason. The phobia developed secondarily as the anxiety is attached to the immediate environment prevailing when the internal anxiety happens to surge. First comes the panic without relationship to the surroundings, the surroundings are then attached or conditioned to the anxiety and agoraphobia has begun.

In the case of simple phobias, learning theory seems to explain at least some aspects of phobia. A small child falls down the stairs and later is afraid of heights. Regarding the most devastating form of phobia, agoraphobia, learning theory seems to contribute little.

Finally, why do more women than men have phobias? Before puberty the distribution of phobia between the sexes is about equal. Later, women are more phobia-prone. Here are several explanations, none satisfactory: (1) men are more physically aggressive than women and aggressive action is incompatible with phobic fear; (2) the culture *expects* men to be brave and therefore men are brave; (3) men really are not braver than women but the culture teaches them to lie about fear. The third point may be the most valid. The only evidence that women have more phobias than men consists of more women *saying* they have phobias.

In any event, the difference between the sexes is not great, and of all the questions about phobia which needs to be answered, this is probably the least pressing.

9

The treatment of phobias

> One can hardly ever master a phobia if one waits till the patient lets the analysis influence him to give it up . . . One succeeds only when one can induce them . . . to go about alone and to struggle with their anxiety while they make the attempt.
>
> Sigmund Freud

Psychiatrists and psychologists have been treating phobias for the past hundred years but until recently there was little evidence the treatments were effective. Traditional psychotherapy — ranging from 'supportive' therapy to psychoanalysis — did not seem to help much, whether the patients were seen individually or in groups. One study found that patients psychoanalysed for 10 or 15 years showed slight or no improvement. Another found that only 13 per cent of a mixed group of phobics were free of symptoms after intensive psychotherapy. (The recovery rate for untreated patients is usually higher than this.) Judging by one study, hypnosis may be more effective than psychotherapy: 37 per cent of patients with a flying phobia recovered after *one* session. Studies showed that phobics in long-term psychotherapy were more co-operative in taking tranquillizers than patients in short-term psychotherapy, but otherwise not much could be said for psychotherapy of any length.

About 25 years ago, this all changed. The watershed year was 1958, when Joseph Wolpe proposed a new treatment for phobia in a book called *Psychotherapy by reciprocal inhibition*. It was not entirely new. Behind the treatment was an old theory — learning theory — dating back to Pavlov and the behaviourists of the 1920s. Its message was not new either. Years before, Freud had said that psychoanalysis was

Phobia: the facts

futile unless phobics 'struggle with their anxiety' and confront what they fear. It *was* new in that Wolpe said it did not matter how the phobia came about, or whether the patient understood the origins of the phobia. The phobia could be attacked head-on as a learned condition that could be unlearned. He gave specific instructions for the unlearning part. He also claimed great success for his treatment: a 95 per cent recovery rate!

Later there was some disillusionment about the Wolpean 'behaviour therapy'; nobody, including Wolpe, ever got so high a recovery rate again. Enthusiasm, however, for other kinds of behaviour therapy has grown steadily. Imaginative, optimistic, always trying something different, a generation of behaviour therapists has become the dominant force in the treatment of phobias. Today, the person who seeks help for a phobia will probably receive some type of behaviour therapy.

It may be with or without drugs. There is rather convincing evidence that two types of drugs usually prescribed for depression relieve panic attacks. Many therapists treat phobias with behaviour therapy *and* drugs.

For these reasons, most of what follows concerns behaviour therapy and drugs.

Behaviour therapy

There are many kinds of behaviour therapy, but they all have a common goal: reduction of anxiety by exposure to the phobic situation. They differ only in how this is accomplished.

The exposure can be 'graded' or 'ungraded'. It can occur in real life or in the person's imagination.

Graded exposure is based on the artichoke principle: you take it leaf by leaf. It is also the behaviourist equivalent of painless dentistry. The only distress the patient may feel is boredom; graded-exposure behaviour therapy *à la* Wolpe can be excruciatingly tedious, for the therapist if not the patient

The treatment of phobias

(probably explaining why so many therapists favour other procedures).

Here is Wolpe's approach:

1. Identify the phobia.
2. Grade the phobia in its various aspects from least fearful to most fearful (called 'constructing a hierarchy').
3. Instruct the patient to imagine, or visualize, a situation involving the least fearful aspect of the phobia.
4. Combine the visualization with a pleasant experience — usually muscle relaxation — to counteract the mild fear created.
5. When the patient can visualize a mildly fearful situation without feeling anxious, 'move up' the hierarchy to a slightly more anxiety-provoking imaginary situation. Continue this, step by step, until the patient can visualize the most fear-arousing situation in the hierarchy and not experience fear.

At this point is he cured? Perhaps not altogether. Neo-Wolpeans insist he must now go out into the world and practise confronting real-life phobic situations for the treatment to be fully effective.

Here is a typical hierarchy for a snake phobia, rating anxiety on a 6-point scale (6 being most anxious).

1. Seeing a picture of a snake: 1-plus.
2. Twenty feet from a small garden snake: 2-plus.
3. Twenty feet from a boa constrictor: 3-plus.
4. Five feet from any snake: 4-plus.
5. Touching a snake with a gloved hand: 5-plus.
6. Holding snake in lap: 6-plus.

The patient is first taught a method of 'deep relaxation' — a pleasurable state. He then relaxes while visualizing a picture of a snake. If no anxiety occurs, he holds up a finger, and the therapist tells him to visualize step 2. No anxiety? Step 3. No anxiety? Step 4. Finally the patient visualizes a

Phobia: the facts

snake in his lap with no anxiety — and no phobia.

Whenever he moves up a step, relaxed but feeling anxious, he is told to go back down a step and try again. Many exhausting sessions may be required to reach step 6, but infinitely fewer than in psychoanalysis — at a fraction of the cost.

The process of moving up the hierarchy is called 'systematic densensitization'. The method works, Wolpe believed, because of 'reciprocal inhibition', referring to the fact that two incompatible emotions cannot be experienced simultaneously.

Two such emotions are fear and pleasure. Relaxation is pleasurable. There are various methods for becoming relaxed. Wolpe favoured the method of Edmund Jacobsen, consisting of first tensing a muscle group and then releasing the tension beyond the normal resting tension. When all muscle groups have been tensed and relaxed, a pleasurable state of deep relaxation results.

Once a person learns to relax deeply, he can then expose himself to mildly fearful stimuli — in his imagination or real life — and not experience the incompatible emotion of fear. Having conquered mild fear, he can practise relaxing in situations involving increasingly greater fear. The sweetly-smiling, drowsy-looking man you saw on the park bench holding the boa constrictor in his lap is likely the satisfied customer of a systematic densensitizer.

There are other pleasures than relaxation. Sometimes a tranquillizer or sedative like sodium amytal is given to phobics moving up a heirarchy. Sometimes relaxation is helped along by hypnosis.

Success with the Wolpe method obviously requires a patient with a vivid imagination. Not everybody has one. Some patients are too anxious to visualize a scary scene. Some find the whole business a bit silly and burst out laughing when told to visualize a boa constrictor. This annoys the therapist and the treatment comes to a halt, phobia intact.

This — plus some doubts about the effectiveness of the

The treatment of phobias

method — led behaviour therapists to try something more daring: expose the patient to *real* snakes (or whatever he fears). Again, a hierarchy is constructed. The patient moves up the hierarchy at his own pace, keeping anxiety at a comfortable level while gradually increasing exposure to the feared object or situation. The difference is that the snakes, the crowds, and the elevators are all real.

Real-life exposure has two drawbacks: (1) some patients refuse to co-operate because of anticipatory anxiety, and (2) real-life hierarchies are hard to construct in real life. Snakes are fairly easily obtained, although perhaps not boa constrictors. Fire escapes are handy for height phobias — steps 1, 2, and 3 become literally steps 1, 2 and 3. But what about thunderstorm phobias? Any loud, booming noise might serve for steps 1, or 2, but a thunderstorm is needed for step 6. Scheduling a therapy session to coincide with a thunderstorm calls for skills beyond those usually possessed by behaviour therapists.

Which raises the question: does the therapist have to be present while the patient moves up a hierarchy? The answer is: it helps. As you may recall from earlier chapters, no *soteria* — fear dispeller — is more effective than a trusted companion. A *paid* trusted companion is perhaps even better, and so therapists take full advantage of their *soteria* role by actively engaging the patient in something called 'participant modelling'.

This wrinkle on Wolpean therapy is based on the principle that nothing succeeds like success — even someone else's success. Before the patient moves up a step on the hierarchy, he watches the therapist demonstrate that it can be done. The therapist moves closer to the snake; the patient then moves closer. The therapist touches the snake; the patient touches it. Courage is contagious and seeing others do something we fear makes it easier for us to do it ourselves. Learning through others — vicarious learning — shapes the raw genetic material

Phobia: the facts

with which we begin life into the finished product we call personality.

The therapist helps the phobic patient climb the hierarchy in various other ways. The patient may be instructed to place his hand on the therapist's hand while the latter touches the snake. The therapist may lavishly compliment the patient for every small success while withholding criticism. He may keep a log of the patient's progress: the length of exposure, the closeness of contact, the number of hierarchy items completed. Whenever the patient exceeds the day's goal for items, he is told, 'That's great!' (Therapists call this 'reinforced practice'.) He may improvise for each patient a number of 'performance aids' — for example, in animal phobias, protective clothing for the patient and a leash for the animal.

Over-involvement by the therapist poses risks. The patient may do well as long as the therapist is there but be as phobic as ever when the therapist is absent. The goal of behaviour therapy, like all therapy, is to eliminate the therapist. All therapies must be tested by the standard: does the new learning transfer to real life? This is why therapists emphasize 'mastery practice' between sessions.

So much for the graded approach to phobias. It works sometimes and sometimes does not. It works best with simple limited phobias. It is usually found somewhat superior to other psychotherapies, but not always. (Most people with snake phobias do not seek psychotherapy.) At least the pioneers of behaviour therapy have been scientifically minded. They have sought proof that their treatments work. Single case histories have been played down. As a result, there exists a body of evidence bearing on the effectiveness of the various behaviour treatments.

One drawback in these studies is that most have been conducted with non-patients — typically college girls with snake phobias or fear of heights. These are problems which in themselves rarely lead to psychiatric consultation.

The treatment of phobias

In any case, it is clear that systematic desensitization, whether in real life or in the person's imagination, was not the panacea its originators had hoped it would be. Behaviourists looked for other approaches and the one that has caught on most in recent years was something called 'flooding'. This involves *ungraded exposure* to the phobic stimulus.

The person is afraid of heights? Take him to the roof of the Empire State building and have him lean over the guard rail. He will be terrified of course. Terror is not an emotion that can be sustained indefinitely and, little by little, he will find his anxiety diminishing. In other words, move directly to step 6 and be done with it. 'People don't die of anxiety', flooders say, then warn against flooding anyone with heart disease or other illnesses where anxiety possibly *can* kill.

The treatment is beautifully simple. A phobia can be cured in a single session, although usually it takes several. The trick is to keep the phobic in a state of maximal panic for as long as necessary for the panic level to drop. If the person is removed from the phobic situation before the panic drops, he will just be worse for the experience. How do you get people to walk into burning houses when they are afraid of fire? Jump off bridges when they are afraid of both heights *and* water? Flooders answer that you just do it. You stay with the patient. You encourage him, you praise him, remind him that he promised he would go through it. You do not resort to physical means but to almost everything else.

Repeated trials of prolonged exposure to real-life stimuli of maximal phobic intensity do indeed produce results, as would have been predicted by learning theory which holds that phobic anxiety is sustained by repeated avoidance of anxiety-eliciting stimuli. The trick is to find people who are willing to go through it. Flooding works with animals: a strongly conditioned fear response can be extinguished by physically restraining the animal in a fear-eliciting situation. People are not so restrainable.

Then, too, there is the problem of eliminating the therapist.

Phobia: the facts

Everyone agrees that phobics cannot go through life with a therapist at their side, or even a friend or relative. The cure of phobia comes when the phobic can confront the feared stimulation and 'go about it alone', as Freud said.

The good therapist actively works at self-elimination.

A 22-year-old man refused to ride a bus by himself. The therapist agreed to accompany him during one session but suggested they sit in different sections of the bus rather than next to each other. The patient agreed to the compromise and was able to tolerate the anxiety elicited during the one-hour bus trip. In the following session, the patient successfully traveled by bus alone.

Real-life flooding is often hard to arrange and therapists fall back on imaginary flooding. After learning what frightens the patient the most, the therapist asks the patient to visualize his most frightening scene and helps by describing the scene. The therapist watches for signs of anxiety in the patient and repeats the scenes which produce the most anxiety. As with real-life flooding, he keeps it up until the anxiety level falls.

Some therapists embellish flooding by describing exaggerated catastrophic scenes to the patient. The patient is afraid to drive and is asked to imagine that he is driving a school bus that plunges over a cliff with everyone killed. Frightened of speaking out at a small meeting, he imagines that he is the keynote speaker at a convention in Madison Square Garden. This is called *implosion therapy*. It is not widely used, being as disliked by most therapists as by patients.

What works best? There are studies showing that flooding is better than systematic desensitization; studies showing that systematic desensitization is better than flooding; and studies showing both with the same effect or no effect. What almost all studies do show is that real-life exposure is superior to imaginary exposure, whether graded or ungraded. The combination of real-life exposure with 'participant modelling' and 'reinforced practice' seems particularly effective.

* * *

The treatment of phobias

Wolpe and his followers have been critical of psychoanalysis, one reason being the emphasis psychoanalysts place on unconscious thought processes. Conditioning therapy, however, also relies on unconscious processes. Conditioned responses are just as 'unconscious' as formation of urine by the kidneys or an Oedipal complex. People do not 'choose' to be conditioned: it just happens. As it is happening, the conscious mind may observe what is going on but is relatively powerless to do anything about it. When a person overcomes a phobia through behaviour therapy, it occurs automatically. Whether he confronts the phobic situation by degrees or all at once, the essence of the cure is deconditioning, which has hardly anything to do with thinking or reasoning.

For the behaviourists the phobic patient is a victim of immutable principles of conditioning, just as Freud's patients are victims of childhood drives, traumas, and frustrations.

Not all behaviour therapists, however, reject conscious thoughts as irrelevant to the acquisition of phobias or their cure. A branch of behaviour therapy called 'cognitive therapy' holds that thoughts cause feelings as well as feelings cause thoughts (a 'cognition' is a fancy word for a thought). Cognitive therapists teach their patients to substitute positive thoughts for negative thoughts. The patent for this approach no doubt belongs to Norman Vincent Peale, but cognitive therapists have approached positive thinking more systematically than did Dr. Peale, and have even done studies, sometimes showing that their treatment works.

As applied to phobias, the treatment consists of training in 'self-statements'. Because people are always talking to themselves (cognitive therapy rests on this premise), why not train people to say things to themselves which make them feel better and improve their life adjustment?

First the patient must audit his thinking and identify those negative thoughts which make matters worse. 'Oh, I'm going to the grocery store today. The place will be full of people. I

Phobia: the facts

may faint! I won't go to the store! I'm sick, I'm weak, I have *agoraphobia*!' The cognitive therapist says, 'Stop! Don't think those things.' 'But I *have* to think those things', protests the patient. 'You don't', responds the therapist. 'You *can* think other thoughts, and I can train you to do so.'

The patient is given some specimens of positive thoughts and encouraged to develop more of his own. 'Okay, so I faint. What difference does it make? I have never fainted before. I am strong and healthy. I'll bet I am stronger and healthier than most of the people in that store. I'm going to the store, and to hell with it!'

The first hurdle in self-statement training is getting the patient to accept the possibility that distressing thoughts can be replaced by comforting thoughts by effort alone. In a sense, resistance to this idea reflects the times. People hear constantly they are victims — victims of the economy, their upbringing, hormones, microwaves, etc. — and come to believe it.

Cognitive therapy has been shown in several studies to produce at least modest improvement in depressed mood and social adjustment. This suggests we may not be *total* victims of our pocketbooks, glands, and environment. Many behaviour therapists have added self-statement training to their armamentarium without making any big claims for it but nevertheless convinced that what a person says to himself and about himself make a difference.

It probably does little harm and may be helpful if patients learn to see a humorous side to their phobias. Humour may put the phobia in perspective and is a wholesome response to most life problems. People do not like to be ridiculed, of course, so humour must be handled with delicacy.

A version of cognitive therapy called *paradoxical intention* almost always makes the patient smile, if not laugh uproariously, and is based on a principle known to all psychologists: the harder a person tries to be unafraid, the more

The treatment of phobias

afraid he becomes. Paradoxically, trying to be afraid may reduce fear.

Any reader can apply this principle to himself by saying repeatedly: 'I will *not* think of a candy-striped elephant'. If you are a normal reader (aha, the rub), the harder you try not to think of candy-striped elephants, the more you think of candy-striped elephants.

Here is how this is applied to phobias: the man with a cardiac phobia is instructed by the therapist to have a heart attack. 'Right now!' the therapist says. 'Here in the office. Go ahead, have one!' The patient thinks for awhile, smiles, shrugs, and says, 'Well, it isn't that easy'.

The patient can practise paradoxical intention when alone. If he fears a heart attack whenever his pulse increases, he is instructed to speed his heart up even faster and *try* to have a heart attack in that fashion. Typically the heart rate accelerates for a few seconds and then drops dramatically.

The phobic patient may be asked to list symptoms that bother him most and attempt to produce them deliberately. Most are not under voluntary control (blushing, trembling, etc.). When the patient cannot directly will them to happen, he feels reassured, Whether paradoxical intention produces lasting improvement is not known, having never been studied.

Relaxation is commonly used as an adjunct to behaviour therapy, as is (less commonly) hypnosis. As mentioned, Wolpe favoured the relaxation method of Jacobsen. *Autogenic training* is a more elaborate technique in which relaxation is one part. The treatment is based on the premise that the autonomic nervous system can be controlled by imagination. If, for example, you imagine feelings of warmth in your hand, more blood will flow to your hand, producing warmth. In one study, phobic reactions responded well to autogenic training, but more studies are needed to prove its usefulness.

Phobia: the facts

Case reports
(See Notes for sources)

Few studies bear on the possible effectiveness of psychoanalysis for phobias, but testimonials abound in the professional journals. The following is notable for its brevity and sensational success:

An eight-year-old boy became frightened after reading about an 11-year-old boy who was dying of 'old age', and developed fears of monsters, baths, and the dark. The father, on advice from his own psychoanalyst, interpreted his son's fears: anger was leading the boy to see his father as a monster. He was told that, rather than seeing monsters, he could get angry at his father. Anxiety diminished rapidly and there was no return of fear of monsters, baths, or the dark.

The following story illustrates systematic desensitization of a dog phobia.

Eleven-year-old Yvonne was not doing well in school. She was shy and lacked confidence, but was not intellectually retarded. Her terror of dogs was the reason for referral. Yvonne was initially afraid that the therapist would show her a large dog. She attributed her fear to an incident at two years of age when a visitor threw a large stuffed bulldog into her playpen.

At the initial session, she was told that she could learn to not fear dogs just as she had learned to fear them. A portable radio would be her prize for being able to pet a dog. Muscle relaxation was demonstrated, and she was taught about imagery. During six additional sessions, she continually glanced and fidgeted. Since she was unable to relax, the desensitization technique was modified. A warm, friendly relationship with the therapist was substituted for muscle relaxation as the anxiety inhibitor. Instead of imagery, pictures of dogs were used.

Three books were used, ranked in order of fewest to most pictures of dogs. With the therapist keeping score, Yvonne pointed to every picture of a dog. She was given homework in which she would write a story about a dog.

She was then asked to make up a story about a dog and tape-record it. While playing it, she imagined what was happening and afterward described the images for the therapist. At the end of

The treatment of phobias

another replay, the therapist added more fearful events, which she imagined (for example, a dog on a leash ran off, played with her, licked her, and so on). After each session, she was asked to draw a picture of the dog she imagined. Homework was also hierarchical, in that numbers of dog pictures had to be cut out and put in a scrapbook. Other assignments were sketching or photographing a live dog, visiting a pet shop, and finally petting a dog.

Her phobia disappeared.

Here is how a noise phobia was treated by flooding.

Bill, nine years old, was referred because he feared sudden, loud noises — bursting balloons, cap guns, motorcycles, and pneumatic drills. These fears prevented normal play with peers and siblings. He sulked, felt people were against him, and was self-conscious.

Desensitization was used for 22 sessions, each lasting from 30 minutes to one hour. The hierarchy consisted of firing caps to bursting balloons and hearing cars backfiring. Actually firing a cap gun while relaxing, conversing, and eating candy, was successful in eliminating that fear. Treatment was discontinued, although other fears persisted.

Nine months later, Bill was still worried about his balloon phobia. The therapist explained that treatment would be unpleasant for a short time, but Bill wanted to try it. Fifty inflated balloons were brought into a small room, frightening Bill very much. He refused to burst any, so the therapist burst six, making loud noises. Bill cried, but the therapist continued bursting balloons until Bill did not flinch at all. With much persuasion, Bill pushed balloons against a nail held by the therapist. Steps followed of bursting balloons by hand with one hand over his ear, then bursting balloons with both ears uncovered. In this first session, 220 balloons were burst and, although shaken, he agreed to return the next day. During the second session, 320 balloons were burst, after initial encouraging and prompting. By the end, he burst them enthusiastically.

A two-year follow-up showed no noise phobias and better peer relationships.

Sometimes behavioural techniques have been applied to groups.

A six-year-old girl developed psychosomatic pains and anxiety when anticipating or hearing a bell for a fire drill. Her classmates became distressed and showed undue concern about fire drills. Systematic

Phobia: the facts

desensitization techniques were successfully used with the girl and her classmates. The bell was paired with a pleasure-producing auditory stimulus — tape recordings of children's songs and stories. It is hypothesized that the group's positive responses to the stories aided in the elimination of the girl's phobia.

One more story about a child, showing how ingenuity is important in behaviour therapy:

A 10-year-old boy was excessively afraid of the dark. He felt acute anxiety when his parents left and when they were home he had to be accompanied when entering any darkened room. This fear presumably originated after he viewed a frightening film and had been warned about burglars and kidnappers. A previous therapist had ameliorated some interpersonal difficulties, but the fear of darkness remained.

After a 30-minute conversation, this boy's great interest was revealed in two radio heroes — Superman and Captain Silver. He was asked to imagine that Superman and Captain Silver appointed him as their agent. A detailed story was told by the therapist: while at home, the boy imagined receiving a signal on his wrist radio. He waits for the heroes, alone, in a dimly lit lounge. If anxiety was indicated in any scene, the story was terminated. By the third session, he could imagine himself alone in his darkened bathroom, awaiting Superman's communication.

At the end of treatment, his phobia was gone.

Children probably respond better to behaviour therapy than adults, perhaps because they are less verbal than adults and behaviour therapy lends itself to non-verbal techniques. Also, phobias are often passing things in children and therapists *usually* have good luck in treating passing things.

With adults, phobias are usually not so passing. Sometimes treatment is elaborate and uncertain. Something more than behaviour treatment was needed — and something more was found.

Drugs for phobias

In the 1960s, as behaviour therapy began to flourish, a group of psychiatrists in England and another in New York began

The treatment of phobias

studying the effectiveness of drugs for phobia. Phobics from time immemorial had taken drugs: bromides, chloral hydrate, barbiturates, derivatives of the poppy plant, and hemp, and, of course, alcohol. All relieve anxiety, but briefly; all produce hangover or withdrawal, and all are potentially addictive. Responsible doctors in the twentieth century prescribe them as little as possible. Then, in the 1950s, three new classes of drugs — man-made and new under the sun — came on the market.

One group was for psychoses. Another was for anxiety. The third was for depression.

The new behaviour therapies helped some phobias, but not all; the sickest phobic, the agoraphobic, was particularly unresponsive to behaviour therapy. So the new drugs were tried, and here are the results.

Tranquillizers

Drugs for psychoses are called 'major tranquillizers' or neuroleptics; Thorazine was the first one introduced. There is no evidence these drugs help phobias.

Drugs for anxiety are called 'minor tranquillizers' or anxiolytics, usually referring to the benzodiazepine class of drugs; Librium and Valium are examples. Since phobias are anxiety states *par excellence*, one would expect that if any drug helped phobias, it would be a drug that relieves anxiety, as Librium and Valium indubitably do.

These drugs indeed are widely prescribed for phobias, but, surprisingly, their effectiveness has not been evaluated. They are widely believed ineffective for agoraphobia, but may help simple and social phobias by lowering anticipatory anxiety and giving the 'chemical courage' required to confront a feared situation.

In the case of flying phobias, the airport bar has performed this service through the years. It is not known how many people today board planes fortified with a minor tranquillizer (sometimes combined with alcohol), but one suspects the number is sizeable. One would hope that if a Libriumized phobic boarded enough airplanes, eventually he could board planes anxiety-free *without* Librium. It is not clear this happens. In animal studies, sedatives and alcohol suppress phobia-like avoidance behaviour, but only as long as the animal remains drugged or intoxicated; when the drug is withdrawn, the avoidance behaviour returns. The situation for humans may not be quite so pessimistic. Many people give tranquillizers credit for overcoming stage fright and eventually perform without them (often, of course, keeping a capsule or two in a jacket pocket just in case).

The minor tranquillizers relieve moderate degrees of anxiety but apparently do not prevent panic attacks. In one study, 57 patients with panic attacks had taken more than one-half million doses of minor tranquillizers, yet continued to have attacks. Since panic attacks are frequent in agoraphobia, presumably minor tranquillizers are not useful for this disorder.

Patients with simple and social phobias usually do not need to take minor tranquillizers on a regular daily basis. They are not generally anxious people, and there is evidence that people who are not generally anxious *become* anxious if they take tranquillizers. The tranquillizer should be taken, if at all, shortly before the anxiety-producing situation. Sometimes, as noted, just carrying them in one's pocket relieves anxiety sufficiently for the person to confront the phobic situation, which is what really counts in achieving a lasting recovery from a phobia.

Valium and other minor tranquillizers are the most widely prescribed medications in the world, and, compared to barbiturates and other sedatives, are relatively safe. It is virtually impossible to kill oneself by taking an overdose of the drugs and relatively few people become physically dependent upon them, meaning they take high doses and have serious withdrawal symptoms, like seizures, when they stop taking the drug. There is some evidence that *long-term* use of minor

The treatment of phobias

tranquillizers, even in the doses normally prescribed, result in some withdrawal effects if the drug is suddenly discontinued, but this is not definitely established. The drugs tend to lose their effect after several weeks of taking them daily, and have another disadvantage: some are slowly eliminated from the body. If taken regularly, they accumulate in the body to rather high levels. This may result in the person being constantly drowsy in the daytime, with the implications this has for driving and handling machinery. However, short-acting tranquillizers are now available and are probably preferable for phobias.

For these reasons, the drugs should only be prescribed by physicians who are thoroughly familiar with them, and today this usually means a psychiatrist.

Drugs for depression

Two new groups of drugs for depression were introduced in the 1950s and 1960s. One group consists of something called MAO inhibitors and the others are called tricyclics. Phenelzine is a member of the first group and imipramine belongs to the second. By constantly tinkering with the molecules, drug manufacturers have brought on the market more than 20 drugs in the two groups. There is little evidence that one drug is superior to another in each group, although the side-effects differ.

In 1962, William Sergeant in England reported that 60 patients with anxiety states, including phobias, responded well to MAO inhibitors (sometimes given with a minor tranquillizer). Nearly a dozen studies over the past 20 years have confirmed the finding. They all show that MAO inhibitors — particularly phenelzine — provide some relief for phobias, reducing the anxiety if not the avoidance behaviour. Although the drugs are usually prescribed for depression, depressive symptoms may be absent and the drugs will still alleviate phobic anxiety.

The drugs take about a month to work and then usually

Phobia: the facts

continue to be effective until stopped. When stopped, the phobias almost always return. In other words, in learning theory terms, the 'learned' fear does not become extinguished or 'unlearned' through repeated exposure to the phobic stimulus when the exposure is facilitated by an MAO inhibitor. The almost inevitable relapse when the drug is discontinued provides strong evidence that the drug actually works. On the other hand, MAO inhibitors are potentially dangerous and physicians balk at prescribing them indefinitely.

The patients similarly balk at taking them indefinitely. When given the drug they are also given a long list of foods, drinks, and other medications to be avoided, including aged cheeses, wines, beer, pickled herring, broad bean pods, yogurt, yeast, hayfever pills, and other antidepressant medications. Most patients can do without broad bean pods, but they face life void of cheese and beer with little enthusiasm. Consequently, most doctors do not prescribe an MAO inhibitor until they have tried the tricyclic class of antidepressants first.

It was also in 1962 when Donald Klein and his group in New York reported that a tricyclic antidepressant, imipramine, was an effective treatment for agoraphobia. In studies over the next two decades, they confirmed their original finding and became increasingly specific about the drug's mode of action.

Klein believes that agoraphobia has three parts: (1) spontaneous panic attacks, (2) anticipatory anxiety, and (3) avoidance behaviour. Imipramine abolishes the panic attack but has little effect on the anticipatory anxiety or avoidance behaviour. He recommends combining imipramine with psychotherapy. In Klein's studies, supportive psychotherapy does as well or better than behaviour therapy in preventing anticipatory anxiety and avoidances. ('Supportive' means giving encouragement, reassurance, and sometimes advice, without delving deeper.) Most of Klein's patients who responded to imipramine had a history of receiving psychoanalysis or other intensive psychotherapy, and neither seemed

The treatment of phobias

to help. A combination of imipramine and supportive psychotherapy not only worked but was economical.

Imipramine did not relieve simple phobias. Klein's explanation was that simple phobias are usually not accompanied by a panic attack. Many students of agoraphobia, starting with Freud, believe that a *spontaneous* panic attack is the initial manifestation of agoraphobia, and what agoraphobics fear most, when they stay at home, avoiding travel and crowds, is repetition of the initial panic. The panic is 'spontaneous' in that there is usually no apparent connection with a traumatic event. People with simple phobias more often associate the beginning of their phobia with a trauma. Agoraphobics often have a history of school phobia and 'separation anxiety' associated with real or threatened loss of a parent, but their first panic attack typically comes 'out of the blue'.

Since imipramine relieves the panic attacks of agoraphobia, and since agoraphobics have a history of school phobia, Klein reasoned that imipramine might also help in school phobia. His group did a study and, sure enough, imipramine had a potent effect on school phobia. It is now probably the best treatment for this problem, when persuasion doesn't work.

A high relapse rate follows the discontinuation of MAO inhibitors, and the same is true of imipramine. Even after patients have taken imipramine for six months or longer, one-third or more resume having panic attacks immediately upon cessation of the drug. In short, there is little or no transfer from drug to non-drug state.

This might have been predicted because of a phenomenon called 'state-dependent learning'. People and animals trained to perform tasks under the influence of mind-altering drugs perform well as long as they are under their influence and do poorly when the drug is withdrawn. Their performance is 'state-dependent', the state being one of intoxication on a particular drug. New behaviour learned in a drug state may not transfer to a non-drug state, and this is one of the dis-

Phobia: the facts

advantages of giving mood-altering drugs, particularly drugs with side-effects or long-term hazards.

Imipramine and other tricyclic drugs have many side-effects. They range from dry mouth and blurred vision to increased appetite and impotence. Not everyone experiences the side-effects and they tend to go away as the drug is taken over a period of time. However, phobic patients seem especially sensitive to the side-effects of MAO inhibitors and tricyclic drugs, explaining why there is a high drop-out rate in studies of these drugs. Phobics are particularly bothered by the stimulant effects of imipramine. The drug makes them jittery and causes insomnia, rather like Dexadrine. Other drugs in the tricyclic family have sedative rather than stimulant effects, and there is limited evidence that these drugs may be as effective as imipramine in suppressing panic attacks. Also, adjusting dosage often reduces side-effects.

Klein noted that not only did imipramine suppress panic attacks but improved the 'quality of life' of his patients. They got along better with friends and relatives, perhaps because some phobias are as troublesome for friends and relatives as for the victim. He noticed something observed by the behaviour therapists: patients relieved of phobias do not develop 'substitute' symptoms. Psychoanalytic theory holds that phobias are symptoms of unconscious conflicts, that if one symptom goes away another appears in its place unless the conflicts are resolved. Studies of phobias over the past 20 years tend to refute this idea.

Like behaviour therapy, drug therapy of phobias was ushered in with much enthusiasm. This has waned somewhat over the past 20 years, without evaporating. Drugs help; hardly anybody questions this any longer. But they do not help everyone and drugs alone rarely cure phobias. Klein recommends a combination of drugs and psychotherapy. Others recommend a combination of drugs and behaviour therapy. In any case, a combination of something is usually required, and no combination represents a truly definitive treatment, in the

The treatment of phobias

sense that penicillin is a definitive treatment for syphilis. People tend to get over phobias. The response to 'placebo' in all studies of phobia is remarkable; at least half of patients improve on a sugar pill alone. Good responders to treatment are often the same people likely to recover anyway: people with good general mental health and an illness of short duration.

Still, there has been much progress in the treatment of phobia over the past 20 years and it is not overly optimistic to foresee a time when the cause of phobia will be understood and the treatment as spectacular as penicillin. (At the time of writing, more than 70 new drugs for anxiety and depression are in various stages of development.)

Beta blockers

Any discussion of drugs for phobias should include propranolol, the most widely prescribed so-called beta blocker. Beta blockers are usually prescribed for high blood pressure and irregular heart action. However, they also are a good remedy for trembling hands, racing pulse, and other physical expressions of anxiety.

William James believed that if you could eliminate the physical expressions of anxiety you would also eliminate anxiety (The James–Lange theory; see Chapter 1). Since nobody's pulse races faster than a phobic's in a phobic situation, and no hands tremble more, it is surprising that beta blockers have not been studied for phobias. One reason they have not been is that they are rather tricky drugs with numerous side-effects, and cannot be taken by people with various common diseases. Nevertheless, the drugs are probably as safe as the antidepressants which, to everyone's surprise, relieves panic as well as relieves depression.

As a matter of fact, beta blockers *have* been studied for one type of phobia: stage fright. Some years ago, clinical pharmacologists of the Royal Free Hospital in London hired Wigmore Hall and engaged 24 string players who had histories of stage

Phobia: the facts

fright to perform under the influence of a beta blocker. It worked. According to a trade newspaper, the drug 'dramatically reduced the effects of stage fright without detriment to technical execution. In fact, teachers, performers and critics involved in the study noted significant improvement in accuracy, rhythmic stability, and memory among the propranolol users.' In the same issue, a violinist called attention to an ethical issue: 'Might not use of potent prescription drugs by a performer at an audition give him an unfair edge over the competitor just as it might to the athlete or race horse? Must orchestras be prepared to administer blood and urine tests to audition applicants?'

Why would an orchestral musician have stage fright? It seems to violate a basic principle of learning theory: phobias last only as long as the phobic situation is avoided. How can an orchestral musician avoid orchestras? There is even safety in numbers: if a violinist has not quite mastered a difficult passage in 'Thus spake Zarathustra', his discreet miming of the fingering and the bowing would not be noticed, except perhaps by the colleague in the next chair.

This notion overlooks the fact that although the performance life of most orchestral players is corporate and comparatively free of anxiety, many performers regularly take solo roles within the orchestra as principals or section leaders. A cellist describes the symptoms he and his colleagues may experience: 'That dreaded onset of sweaty palms, racing pulse, trembling hands, dry mouth, labored breathing, nausea and memory loss.'

There are more phobias around than most people suspect. There are also many treatments, with more to come.

Notes and references

Chapter 1

p. 1

Isaac M. Marks, M.D. of the Institute of Psychiatry at the Maudsley Hospital in London is the world's leading authority on phobias. His book *Fears and phobias* (Academic Press, 1969) is the most scholarly book on the subject in the English language. It provided invaluable background material for this chapter as well as the chapters on phobic disorders. Dr Marks subsequently has published widely in the field of fear and phobia. When Marks is mentioned in the text or in these notes, the source of information was *Fears and phobias*. References to his other writings will be given in full each time they appear.

Other books used for background in this chapter were *The biology of anxiety* edited by Mathew (Brunner Mazel, 1982); *Handbook on stress and anxiety* by Kutash and Schlesinger (Jossey-Boss, 1980); *Handbook of studies on anxiety* edited by Burrows and Davies (Elsevier North-Holland Biomedical Press, 1980); *The meaning of fear* by Rachman (Freeman, 1978); *The psychology of anxiety* by Levitt (Erlbaum, 1980); *Anxiety and emotions: Physiological basis and treatment* by Kelly (Thomas, 1980); *Phenomenology and treatment of anxiety*, edited by Fann (Medical and Scientific Books, 1979); *The role of bodily feelings in anxiety* by Tyrer (Oxford, 1976); *The experience of anxiety* by Goldstein and Palmer (Oxford, 1975); *Clinical anxiety* by Lader and Marks (Grune & Stratton, 1971); *Bodily changes in pain, hunger, fear and rage* by Walter Cannon, a classic (Appleton, 1915).

The distinction I make between fear and anxiety is not one found in the dictionary. Freud defined the words as I do, and most psychiatrists, when they think about it, probably do likewise. The distinction applies only to humans. 'Fear', when applied to animal behaviour, refers to a defensive response which resembles human responses associated with the subjective experience of fear. When a pearl fish flees from danger into the anus of a sea cucumber, this is called a fear response on the grounds that, in circumstances where this was the only haven available, a human would probably do the same.

Guilt is a type of fear: the fear you have done something wrong (you may

Phobia: the facts

not know what) and may suffer from it. What is 'wrong' is culture-bound. Huckleberry Finn felt guilty because he helped Nigger Jim escape slavery — and now would feel guilty if he hadn't. Guilt, in our time, is probably the fear most of us experience most often. In *How to be a Jewish mother* by Dan Greenberg (Random House, 1964), mothers are advised to make their sons feel guilty: 'If you don't know what he's done to make you suffer, *he* will'.

Shame is the fear of being seen naked — perhaps not *really* naked (although even stripteasers have unpleasant dreams of being unclothed in public) but naked in the sense that people see us as we are: stupid, inept, unattractive, worthless (and most of us fear we are one or all of these from time to time).

Jealousy and *envy* are fears. The first is a fear of losing what you have (eroticized in the case of spouses and lovers) and the latter is a fear of somebody else being treated better than you are: a favourite child.

These are emotions and what is an emotion? It is perhaps the hardest word of all to define. Seen from outside, emotion suggests a person in a special state: he is 'e-moved', 'moved out of himself', likely to act for a time in a more or less unusual way.

From the inside, emotion is a feeling that prompts quick action of some kind as opposed to a *thought* which judges whether the action is possible, justified, or in one's best interest. Feelings and thoughts are often at odds: the thinker cannot decide what is best and the feeling is mixed, contradictory, a jumble of fear, anger, and other ingredients of an emotional goulash.

These definitions are not in the dictionary either. Definitions, at bottom, are arbitrary. 'When I use a word', Humpty Dumpty told Alice, 'it means just what I choose it to mean — neither more nor less.' The above words mean, more or less, what I choose them to mean, and let those who argue, argue with Humpty Dumpty.

p. 2

Facial expressions: *On the expression of emotions in man and animals* by Charles Darwin (John Murray, 1872).

Translating Yoruba: *Psychiatry around the globe: A transcultural view* by Julian Leff (Dekker, 1981).

The quotation by Burton is from the 11th edition of *The anatomy of melancholy*, published in 1813.

pp. 3–4

Fight and flight are not the only possible responses to danger. Immobility is a third. It is observed often in animals. Rather than running away or fighting

Notes and references

back, the animal 'freezes' — standing absolutely still, rolling in a ball, sometimes pretending to be dead. The seven emergency responses orchestrated by the sympathetic nervous system are replaced by defensive responses controlled by the parasympathetic system. Instead of speeding up, the heart rate becomes slower and may even lead to cardiac arrest and death. This is sometimes called a 'vagal' death, because the vagus nerve — a branch of the parasympathetic system — makes the heart beat slower. What is called 'voodoo death' in humans may involve the same mechanism. Immobility may actually have survival value. There are many reports of predators passing by immobile animals to pursue fleeing ones. When a human becomes 'paralysed with fear', the physiology may be the same as occurs in animals 'playing possum', and there are certainly situations where a Gandhi-like passivity turns away wrath better than fleeing or fighting.

p. 5

Re: Fear centres: An excellent resource is the *Neuropsychology of anxiety: An inquiry into the functions of the septo-hippocampal system* by Jeffrey A. Gray (Oxford, 1982).

One of the great scientific discoveries of recent years was the finding that the brain contains substances which resemble morphine (called endorphins) and which seem, indeed, to suppress pain. They are found not only in human brains but in brains of the lowest animals, including earthworms. Lewis Thomas, America's great physician–essayist, says he understands why humans might need something like morphine in their brains. 'For a species as intelligent and at the same time as interdependent and watchful of each other as ours, it might be useful to install a device of this kind to guard against intolerable pain, or to ease the individual through what might otherwise be an agonizing process of dying. Without it, living in our kind of intimacy, at our close quarters, might be too difficult for us, and we might separate from one another, each trying life on his own, and the species would then, of course, collapse.'

But why, he asks, should mice have the same equipment, and why, of all creatures, earthworms? Then he thinks about it and decides that maybe earthworms need a little morphine in their brains, also. 'Without protection against overwhelming pain, the day-to-day life of a worm, being stepped on, snatched by birds, ground under plows, washed away in streams, would be hellish indeed.'

Anyway, he says, 'There it is [endorphins], a biologically universal act of mercy. I cannot explain it, except to say that I would have put it in had I been around at the very beginning, sitting as a member of a planning committee, say, and charged with the responsibility for organizing for the future a closed ecosystem, crowded with an infinite variety of life on this planet. No such system could possibly operate without pain, and pain receptors would have to be planned in detail for all sentient forms of life, plainly for their

Phobia: the facts

own protection and the avoidance of danger. But not limitless pain; this would have the effect of turbulence, unhinging the whole system in an agony even before it got under way . . . I would have cast a vote for a modulator of pain . . . set with a governor of some kind, to make sure it never got out of hand.' *The youngest science: notes of a medicine-watcher* by Lewis Thomas (Viking, 1983).

p. 7

Re: 'There is no question that animals inherit fears': This is true in a general sense, but one should be wary of attributing behaviour to heredity simply because it is hard to think of another explanation. Consider this: fetuses can hear perfectly well from the uterus. Loud, angry voices *may* induce fear responses in the fetus just as they do in the infant. Conceivably, if the emergency responses of the fetus are activated repeatedly or at critical periods in development, there may be lasting effects on the nervous system. Is it possible that one child is a bully, and another timid, and a third hyperactive because of overheard conversations during fetal life? There is not a shred of evidence for this, but, yes, it is possible.

To repeat: not everything that seems inherited is really inherited. The disease kuru is a fine example. It is a disease of the nervous system that resembles multiple sclerosis and occurs only in natives of a small South Pacific island. It occurs in middle life, is almost always fatal, and runs in families. All the scientists agreed that it was an inherited disease — until, one day, someone suggested an environmental explanation. It turns out these particular natives are cannibals who eat only the brains of close relatives. If these brains contain a microorganism called a 'slow virus', now believed to cause kuru, then the disease would be passed on from generation to generation through dietary habits and not genes.

The 'hawk effect' is one of the oldest observations in ethology, being first described more than 100 years ago. Konrad Lorenz wrote about it in the early 1950s and the description given here comes from the book by Adam Smith called *Powers of mind* (Random House, 1975).

The first experimental studies of the hawk effect were conducted in the late 1950s at the Regent's Park Zoo (London) by Melzak, Penick, and Beckett (*Journal of Comparative and Physiological Psychology* **42**, 694–8) Dr Penick, now a Professor at Kansas University, recalls the studies as 'fun' despite 'the soggy British weather, the interminable tramway rides to Regent's Park Zoo, and the double-vision caused by intently observing the antics of small feathered creatures as cut-out shapes whirled above their heads.' (She has the office next to mine, which is how I know these things.)

Notes and references

p. 8

A fear of strangers clearly has survival value. The term 'child abuse' conjures up images of cruel parents, but actually children are more likely to be abused by strangers — or killed — than by a mother whose 'maternal instinct' almost always, fortunately, overrides her homicidal impulses.

The observations by Hebb are from *A textbook of psychology* (W.B. Saunders, 1964).

p. 10

The quotation from Stanley Hall appeared in the *American Journal of Psychology* **8**, 147–249.

The LSD user was Adam Smith.

p. 12

The mixed-up raven is found in *King Solomon's ring* by Konrad Lorenz (Methuen, 1952).

The Hebb quote is from *A textbook of psychology* (W.B. Saunders, 1964).

p. 17

The Mark Keller quote is cited in *Alcoholism; The facts* by Goodwin (Oxford, 1981).

p. 21

The wonderfully iconoclastic comment on gossip was made by Dan Davin in *Closing times* (Oxford, 1975).

p. 22

Music is a soteria *par excellence*:
>And so I sing,
>As the boy goes by
>The burying ground,
>Because I am afraid
>*Emily Dickinson*

Phobia: the facts

Chapter 2

p. 25

The phobic psychiatrist is found in the chapter on phobic disorder by John C. Nemiah in *Comprehensive textbook of psychiatry* Third edition, edited by Kaplan, Freedman, and Sadock (Williams & Wilkins, 1980).

p. 26

Errera and Coleman estimated that 20 per cent of psychiatric patients had phobias (*Journal of Nervous and Mental Diseases* **136**, 267-71). Other estimates run as high as 44 per cent. Errera, however, in an excellent review of historical aspects of phobia, estimated that only 2 per cent of patients ever see psychiatrists for *treatment* of phobics (*The Psychiatric Quarterly* **36**, 325-36).

The study comparing patients with predominant mood disorders and patients from a fracture clinic was conducted by Shapira, Kerr, and Roth (*British Journal of Psychiatry* **117**, 25-32).

The unexpectedly high rate of agoraphobia and social phobia in alcoholics was reported by Mullaney and Trippett in the *British Journal of Psychiatry* **135**, 565-73. Terhune also reported that phobics were susceptible to alcohol and drug dependence (*Archives of Neurology and Psychiatry* **62**, 162-72), but Sim and Houghton found a low rate of alcohol and drug dependence in phobics (*Journal of Nervous and Mental Diseases* **143**, 484-91), leaving the issue unresolved.

The Vermont study was by Agras, Sylwester, and Oliveau, 'The epidemiology of common fears and phobia', *Comprehensive Psychiatry* **10**, 151-6.

p. 28

Re: Hydrophobia: 'When too little has been done for such a wound (bite of a mad animal), it usually gives rise to a fear of water which the Greeks call *hydrophobia*.
    . . . In these cases there is very little hope for the sufferer. But still there is just one remedy, to throw the patient unawares into a water tank which he has not seen beforehand. If he cannot swim, let him sink under and drink, then lift him out; if he can swim, push him under at intervals so that he drinks his fill of water even against his will; for so his thirst and dread of water are removed at the same time. Yet this procedure incurs a further danger, that a spasm of sinews, provoked by the cold water, may carry off a

Notes and references

weakened body. Lest this should happen, he must be taken straight from the tank and plunged into a bath of hot oil.'
Celsus (25 BC–AD 50) in De Medicina,
Volume 27, translated by W.G. Spencer

Even in this first description of phobia, direct confrontation with the phobic situation was considered the surest cure. Unfortunately, the patient still died, rabies being fatal.

Hippocrates: *On Epidemics*, V Section LXXXII.

Westphal's classical paper on agoraphobia appeared in the *Archives für Psychiatrie und Nervenkrankheiten* **3**, 138–71 (1871).

p. 29

Re: 'This trend': For the official American classification of psychiatric disorders, see *The Diagnostic and Statistical Manual of Mental Disorders* (DSM-III), third edition, published by the American Psychiatric Association in 1980.

In his 1970 paper, 'The classification of phobic disorders', Isaac Marks discusses the historical problems in classifying phobic disorders (*British Journal of Psychiatry* **116**, 375–86).

Kraepelin, E. *Lectures on clinical psychiatry*, second edition. Thieme, Leipzig (1913).

Freud, S. 'The justification for detaching from neurasthenia a particular syndrome: the anxiety neurosis'. In *Collected works* Vol. 1. Hogarth Press (1894).

Chapter 3

p. 30

Aeroplane phobics have devised ingenious ways to conquer their phobia. One man, for example, learned that the probability of a bomb being hidden in an aeroplane was one in ten million. The probability of *two* bombs being hidden in an aeroplane was one in a billion. To cure his phobia he always carried a bomb with him on aeroplanes.

Recommended reading for this chapter is *Phobia: psychological and pharmacological treatment*, edited by Mavissakalian and Barlow (Guilford Press, 1981) and, of course, the classical work of Marks described in the Notes for Chapter 1.

Other books used for background in this chapter included *Phobias and obsessions* by Melville (McCann & Geoghegan, 1977) and *Phobias: their nature and control* by Rachman (Thomas, 1968).

Phobia: the facts

p. 31

The groundbreaking book by Joseph Wolpe was called *Psychotherapy by reciprocal inhibition* (Stanford University Press, 1958).

p. 32

One reason for so many fancy names for phobias is that doctors have long recognized that fancy names are comforting and even therapeutic. The psychiatrist, E. Fuller Torrey, compared witch doctors with psychiatrists. Both, he said, make you feel better because of the certainty exuded by the 'authority figure, the diploma on the wall, or the proper headdress, bones and rattles, and finally, because the authority in each case gave the condition a name' (from *The mind game* (Bantam, 1973)). You have a *curse* from your dead mother-in-law, or you have a *bug* that's going around. Torrey called it the Rumpelstiltskin effect. Rumpelstiltskin was a dwarf who helped a young woman become queen if she would give him her first-born child. When she asked for the child back, he refused because she didn't know his name. If she learned his name in three days, he said, she could keep the child. At midnight on the third day there is a scene where the queen says archly, 'Is your name Michael? John? Rumpelstiltskin?' And poor Rumpelstiltskin goes pop! Vaporizes. Disappears. If you can give it a name, it will disappear (Adam Smith's version in *Powers of mind* (Random House, 1975)).

p. 36

Freud's discussion of animal phobia appeared in *Totem and taboo* (Hogarth Press, 1955).

Why do some people experience fear looking *up* at a high building as well as looking down from one? There has never been a good answer.

pp. 36–7

Variations on the theme of heights and the notion that a fear of heights is really a fear of falling comes from Isaac Marks.

p. 38

MacAlpine has an excellent article on syphilophobia in the *British Journal of Venereal Diseases* **33**, 92–9.

Rogerson describes how a man stopped worrying about having syphilis once he found he actually had it (*British Journal of Venereal Diseases* **27**, 158–9).

Notes and references

p. 39

Ladee has an excellent review on hypochrondriasis called *Hypochrondriacal syndromes* (Elsevier, 1966).

The 1969 paper by Agras, Sylwester, and Oliveau (Chapter 2 Notes) and Isaac Marks present data on the prevalence of illness phobia.

p. 41

The three follow-up studies of phobia are 'Phobic disorders four years after treatment: a perspective follow-up', by Marks (*British Journal of Psychiatry* 118, 683–8); 'The natural history of phobia' by Agras, Chapin, and Oliveau (*Archives of General Psychiatry* 26, 315–17); 'Treatment of phobias' by Klein, Citron, Woerner, and Ross (*Archives of General Psychiatry* 40, 139–45).

p. 43

Nemiah's remarks on Bunyan's phobia are in *Comprehensive textbook of psychiatry* pp. 1493–1505 (Williams & Wilkins, 1980).

Chapter 4

p. 44

The reader can try a little experiment to demonstrate the power of staring. Stand on a street corner where cars pull up at a street sign. Stare at some drivers and not at others. Those you stare at will move away from the street sign faster than those you ignore. (Sometimes, anyway.)

Those super-polite people, the Japanese, seem to have more than their share of social phobias. These are lumped under the name '*shinkeishitsu*' and include a fear of blushing and eye-to-eye contact; a fear of emitting bad odours; and a fear that one's facial expression may annoy others (Caudill and Schooler, *Mental health research in Asia and Pacific*, East-West Center Press, 1969).

p. 45

Among other early public performances are walking, saying words, doing little dances, playing patty-cake, mimicking others, and fetching things, but these are almost totally ignored by the Freudians in favour of a stupendous emphasis on toilet training. Today, with disposable diapers, one wonders whether soiling is still as unsettling for mothers as perhaps it was in more

Phobia: the facts

messy times. Obsessive-compulsive neurosis was believed by Freud to derive from the 'anal stage of psychosexual development' (foulups in toilet training) but neo-Freudians such as Erik Erikson in *Childhood and society* (Norton, 1950) later associated this stage in development with success or failure in achieving feelings of autonomy and competence.

p. 47

A recent American President was said to dictate to his secretaries (both sexes) while seated on a commode. The President may have been counterphobic (see p. 23).

p. 48

One reason women may have more urinary tract infections than men is that they develop stretched bladders from delaying urination. Many little girls will go to school and not urinate until they return home, sometimes because of a phobia about sitting on strange toilet seats. Another explanation unrelated to phobias is that women take more baths than men and have a shorter distance for bugs to travel to reach the bladder.

p. 50

The standard textbook for treating social-sexual phobias is *The new sex therapy* by Helen Singer Kaplan (Brunner/Mazel, 1974).

p. 52

Hippocrates' description of crowd phobia is found in Robert Burton's *The anatomy of melancholy*.

More about blushing on p. 66.

Nemiah's patient is described in his chapter on phobia in *Comprehensive textbook of psychiatry* (Williams & Wilkins, 1980).

Chapter 5

p. 55

A book by Mathews, Gelder, and Johnston called *Agoraphobia* (Guilford Press, 1981) is the most comprehensive book on the subject. It is heavily referenced and contains a complete description of a self-help treatment

Notes and references

method. Gelder worked with Isaac Marks, whose book, *Fears and phobias*, was also used as a resource for this chapter.

p. 56

Along with *Platzschwindel* and barber's chair syndrome, agoraphobia has a host of synonyms, including phobic-anxiety-depersonalization syndrome, locomotor anxiety, street fear, phobic anxiety state, and non-specific insecurity fears.

Donald Klein and other authorities believe that fear of crowded places and open spaces are really secondary to a dread of being away from home where help cannot be readily obtained in the event a panic attack occurs. Like Freud, they believe that agoraphobia is almost always ushered in by a panic attack which is so frightening that the person wishes fervently to avoid a recurrence, and this leads to the many phobic avoidances which characterize the disorder.

pp. 58-9

Vivid specimens of agoraphobic behaviour can be found in Marks' 1970 paper, 'Agoraphobic syndrome' in the *Archives of General Psychiatry* (23, 538-53); *Agoraphobia in the light of ego psychology* by Weiss (Grune & Stratton, 1964); 'A long-term follow-up study of neurotic phobic patients in a psychiatric clinic' by Errera and Coleman in *The Journal of Nervous and Mental Diseases* 136, 267-71; and 'Confinement in the production of human neuroses: the barber's chair syndrome' in *Behavioral Research and Therapy* 1, 175-83.

p. 61

See Notes on Chapter 2, for studies on phobia and alcoholism. Further evidence that agoraphobia is *not* a variant form of depression was found in a recent study in Iowa. Of the immediate relatives of agoraphobics, 9 per cent had agoraphobia, but there was less depression in the relatives of agoraphobics than in non-psychiatric patients. The study provided further evidence of a connection between agoraphobia and alcoholism: 35 per cent of the fathers of agoraphobics were alcoholic compared to 10 per cent of fathers of non-psychiatric patients. ('A family study of agoraphobia' by Harris, Noyes, Crowe, and Shaudhry, *Archives of General Psychiatry*, in press).

p. 62

Marks and Herst found that one-fifth of agoraphobics in Britain reported

Phobia: the facts

that they had a close relative with the same kind of phobia. ('The Open Door: A Survey of Agoraphobics in Britain', *Social Psychiatry* 5, 16-24.

Chapter 6

p. 64

Three useful books about phobias in childhood are *Treating children's fears and phobias: a behavioral approach* by Morris and Kratochwill (Pergamon Press, 1982); *Anxiety and defensive strategies in childhood and adolescence* by Smith and Danielsson (International University Press, 1982); and *Children's fears* by Thomas (Family Health Publications, 1952).

p. 65

Read *Pain and anxiety control in dentistry*, edited by Stanley R. Spiro (Burguss Press, 1981) if your child has a morbid fear of doctors and dentists. Also worthwhile for doctors and dentists.

Studies of what children are afraid of and when they become afraid are reviewed in Isaac Marks' book *Fears and phobias*, which can be supplemented and brought up-to-date with an article called 'Age-sex trends of phobic and anxiety symptoms in adolescence' by Abe and Masui (*British Journal of Psychiatry* 138, 297-302) and a book by Shepard, Oppenheim, and Mitchell called *Childhood behavior and mental health* (University of London Press, 1971).

p. 66

The Abe and Masui article (see above) described the two phobias which peak in adolescence.

p. 68

Everything you ever wanted to know about Freudian fears (or anything else Freudian) can be found in *The psychoanalytic theory of neurosis* by Otto Fenichel (Norton, 1945).

pp. 69-71

A good paper on school phobia is by L. Hersov (*Journal of Child Psychology* 1, 130-6). Many of the comments about the problem came from this source as well as from Isaac Marks' book.

Notes and references

p. 70

Billy's school phobia is described in an article by R.E. Smith and T.M. Sharpe in *The Journal of Consulting and Clinical Psychology* **35**, 239–43.

p. 71

Reluctant mothers are well described in Eisenberg's paper on school phobia in *The American Journal of Psychiatry* **114**, 712–18.

p. 72

Seventy-one school-phobic children seen at a child guidance clinic were followed up after they passed school-leaving age. Most were doing well, ('School phobic children at work', by Hazel Baker and Ursula Wills, *British Journal of Psychiatry* **135**, 561–4).

p. 73

Donald Klein and his wife, Rachel Gittelman-Klein, found that imipramine, an antidepressant drug, was useful in relieving panic attacks in patients with agoraphobia. They proceeded to see whether the drug also relieved symptoms of school phobia, and it did. ('School phobia: diagnostic considerations in the light of imipramine effects', *Journal of Nervous and Mental Diseases* **156**, 199–215).

Chapter 7

p. 75

Synonyms for panic disorder include neurasthenia, neurocirculatory asthenia, DaCosta's syndrome, effort syndrome, irritable heart, soldier's heart, anxiety reaction, and vasomotor neurosis.

More detailed descriptions of the three psychiatric disorders discussed in Chapter 7 can be found in *Psychiatric diagnosis* by Goodwin and Guze (Oxford, 1979) and *Comprehensive textbook of psychiatry*, edited by Kaplan, Freedman, and Sadock (Williams & Wilkins, 1980).

> O lift me from the grass!
> I die! I faint! I fail!
> My cheek is cold and white, alas!
> My heart beats loud and fast.
> *Percy Bysshe Shelley*

Phobia: the facts

Was Shelley having a panic attack? If so, he probably would have called it something else. A good possibility is 'vapours', a favourite expression in the nineteenth century for anxiety reactions. Vapours also described fainting. Anxiety neurotics sometimes faint — probably from hyperventilating — and in the nineteenth century fainting among women was fashionable. Historians tell us that in Victorian times the prototype of a refined young woman was a 'swooner, pale and trembling, who responded to unpleasant or unusual social situations by taking to the floor in a graceful and delicious maneuver, in no way resembling the crash of the epileptic'. A Jane Austen heroine found one social situation 'too pathetic for the feelings of Sophie and myself. We fainted alternately on a sofa'. Overtight corsets may have been responsible for some of the fainting. A nineteenth-century physician, Dr John Brown, cured fainting by 'cutting the stay laces, which ran before the knife and cracked like a bow string'.

p. 76

A Boston cardiologist, Paul Dudley White, and colleagues published the first follow-up study of panic disorder, the basis for much of what is known about the course and outcome of the illness ('Neurocirculatory asthenia' by Wheeler, White, Ried, and Cohen, *Journal of the American Medical Association* **142**, 878-9).

p. 77

A good description of mitral valve prolapse appears in *Circulation* **54**, 3-20, by Devereux *et al*.

Synonyms for obsessive–compulsive disorder are obsessional neurosis, obsessional state, obsessional–ruminative state, phobic–ruminative state, psychasthenia.

pp. 77–83

For a review of studies of obsessive–compulsive disorder see 'Follow-up studies in obsessional neurosis' by Goodwin, Guze, and Robins (*Archives of General Psychiatry* **20**, 182–7).

p. 78

The Jaspers quote is from his monumental *General psychopathology* (Manchester University Press, 1963).

Notes and references

p. 83

Melancholia is perhaps the oldest term for depression. Schizoaffective schizophrenia (American term) is now believed by many authorities to be a variation of depression, as is psychogenic psychosis and cycloid psychosis (European terms).

A 'clinical depression' primarily involves a loss of interest and pleasure in interesting and pleasurable things: food, sex, hobbies, and life itself, so that clinical depression is associated with suicide, whereas fear is not ('clinical' means severe, prolonged, often requiring treatment).

Sadness and happiness have a curious relationship to time. A character in a book by the novelist, Anne Tyler, comments:

Everything comes down to time in the end — to the passing of time, to changing. Anything that makes you happy or sad, isn't it all based on minutes going by? Isn't happiness expecting something time is going to bring you? Isn't sadness wishing time back again? Even *big* things — even mourning a death: aren't you really just wishing to have the time back when that person was alive? Or photos — ever notice old photographs? How wistful they make you feel? Long-ago people smiling, a child who would be an old lady now, a cat that died, a flowering plant that's long since withered away and the pot itself broken or misplaced . . . isn't it just that time for once is stopped that makes you wistful? If only you could turn it back again, you think. If only you could change this or that, undo what you have done, if only you could roll the minutes the other way, for once.

p. 84

Not all studies show a high prevalence of depression in the families of agoraphobics (Notes, Chapter 5). Present evidence indicates that depression and agoraphobia are two separate disorders.

p. 87

The suicide rate actually has risen, notably among young men, in America since the 1950s. To some extent this may be the fault of physicians. Many people who commit suicide see their doctors within a few months before their deaths with depressive symptoms and even talking about suicide. Too often, depressed patients are treated with inadequate doses of antidepressant medications or are denied electroconvulsive therapy after drugs have failed to help.

Guns are the main source of suicide by men in the US.

In Britain the suicide rate has declined during the past 20 years, one reason apparently being that the carbon monoxide content of gas used in homes was lowered, thus making less lethal a once favourite means of exit for depressed Britons ('Suicide in Britain: more attempts, fewer deaths, lessons for public policy' by J.H. Brown (*Archives of General Psychiatry*, **36**, 1119-24); 'The increasing rate of suicide by firearms' by J.H. Boyd (*New England Journal of Medicine* **308**, 872-4)).

Phobia: the facts

p. 88

Evidence of the effectiveness of electroconvulsive therapy can be found in 'Electric convulsion therapy in depression: a double-blind controlled study' by Eric D. West (*British Medical Journal* **282**, 355-7) and the 'Northwick Park electroconvulsive therapy trial' by Johnstone *et al.* (*Lancet*, 20 December, 1980).

Chapter 8

p. 90

The story of Little Hans can be found in Freud's *Collected works* Vol. 3, pp. 149-89 (Hogarth Press, 1909).

p. 92

The case histories are from 'Psychotherapy and phobias' by J.D.W. Andrews in *Psychological Bulletin* **66**, 455-80.

pp. 92-3

Freud's gift for finding sexual significance in street traffic and shopping is developed by R.P. Snaith in 'A clinical investigation of phobias' in the *British Journal of Psychiatry* **114**, 673-7.

p. 93

The philosopher Karl Popper eloquently defends the idea that science can only prove that something is untrue and never that something is true, and that 'falsifiability' is the test of a good scientific theory.

p. 94

The story of Little Albert was first described in a 1920 paper in the *Journal of Experimental Psychology* (**3**, 1-14) called 'Conditioned emotional reactions'.

p. 95

B.F. Skinner coined the term 'operant conditioning'. The real father of operant conditioning, however, was A.H. Thorndike, who formulated the 'law of effect' in his Ph.D thesis in 1898.

Notes and references

p. 99

Wolpe reported the case in *Psychotherapy by reciprocal inhibition* (Stanford University Press, 1958).

Chapter 9

p. 105

Two excellent recent books on the treatment of phobia are *Phobia* edited by Mavissakalian and Barlow (Guilford Press, 1981) and *Agoraphobia* by Mathews, Gelder, and Johnston (Guilford Press, 1981). Also recommended are *Phobia: a comprehensive summary of modern treatments* by DuPont (Brunner/Mazel, 1982); *The behavioral management of anxiety, depression and pain* edited by Davidson (Brunner/Mazel, 1976); and *Anxiety: its components, development, and treatment* by Lesse (Grune & Stratton, 1970).

Books on fear and phobia written for the lay reader include *Living with fear: understanding and coping with anxiety* by Marks (McGraw-Hill, 1978); *Phobias and coming to terms with them* by Emerson (British Medical Association, 1969); and *How to master your fears* by Steincrohn (Funk, 1952).

Wolpe's book was published by Stanford Press in 1958. Evidence that reciprocal inhibition is effective for simple phobias is presented in a paper by McCononaghy in the *British Journal of Psychiatry* **117**, 89–92 and reviewed in the book edited by Mavissakalian and Barlow (see above).

p. 108

Jacobson's book on progressive relaxation was published by the University of Chicago Press in 1938.

A more recent book by Jacobson is *Anxiety and tension control* (Lippincott, 1964).

p. 113

Cognition actually derives from the Latin *cognito*, meaning to learn or acquire knowledge.

Aaron Beck, the father of cognitive therapy, describes the technique in *Cognitive therapy and the emotional disorders* (International Universities Press, 1976).

Phobia: the facts

p. 114

Studies showing that cognitive therapy relieves depression include one by Rush *et al.* in *Cognitive therapy and research* **1**, 17-37 and another by Shaw in the *Journal of Consulting and Clinical Psychology* **45**, 543-51.

Hilarity as a treatment for phobia was first described by Frankl in 'Paradoxical intention: a logotherapeutic technique', *American Journal of Psychotherapy* **14**, 520-35.

p. 115

Autogenic training is described in a book by that name by J.H. Schultz and W. Luth (Grune & Stratton, 1959).

pp. 116-18

An eight-year-old boy: Mintz, I.L. 'Fleeting phobias', *Journal of the American Academy of Child Psychiatry* **9**, 394-5.

Eleven-year-old Yvonne: Kissel, S. 'Systematic desensitization therapy with children: a case study and some suggested modifications', *Professional Psychology* **3**, 164-8.

Bill, nine-years-old: Yule, W., Sacks, B., and Hersov, L. 'Successful flooding treatment of a noise phobia in an eleven-year-old', *Journal of Behavior Therapy and Experimental Psychiatry*, **5**, 209-11.

A six-year-old girl: O'Reilly, P.P. 'Desensitization of fire bell phobia', *Journal of School Psychology* **9**, 55-7.

A ten-year-old boy: Lazarus, A.A. and Abramovitz, A. 'The use of "emotive imagery" in the treatment of children's phobias', *Journal of Mental Science* **108**, 191-5.

pp. 120-1

The strongest evidence that long-term use of benzodiazepine minor tranquillizers results in withdrawal effects is presented in a paper by Malcolm Lader in the *Journal of Clinical Psychiatry*, April, 1983. The most common withdrawal symptoms were anxiety, tension, and sleep disturbance, but a small minority of patients also had more serious withdrawal effects, including paranoid reactions and visual hallucinations. The withdrawal symptoms usually went away within two to four weeks. Lader also reported data suggesting that tolerance does not develop to therapeutic doses of

Notes and references

benzodiazepines (i.e. the drugs do not lose their effects even after several weeks of taking them daily). Other reports, however, report development of tolerance to these drugs and a low incidence of withdrawal effects.

p. 121

Studies indicating that MAO inhibitors relieve anxiety states, including phobias, are reviewed by Grunhaus, Gloger, and Weisstub in *The Journal of Nervous and Mental Disease* **169**, 608-13.

p. 122

Two articles appeared in the February, 1983, issue of the *Archives of General Psychiatry* presenting the latest evidence that drugs relieve certain aspects of phobias. The two articles, however, provide somewhat different conclusions. A New York group headed by Donald Klein reported that imipramine prevented panic attacks in phobic patients, although it did not relieve anticipatory anxiety. (Klein elsewhere recommends benzodiazepines for anticipatory anxiety.) Klein also reported that supportive psychotherapy (consisting mainly of reassurance) was as effective as behaviour therapy, as long as both led to confrontation with the phobic situation. In contrast, a British group headed by Isaac Marks concluded that imipramine had no effect on panic attacks, but 'in vivo self-exposure' was a potent treatment for agoraphobia. The Klein group gave higher doses of imipramine and studied the results while the patients were still taking the drug, whereas the Marks group evaluated the drug effects after the drug had been terminated. As mentioned in this chapter, drugs only seem to be useful in relieving phobia while they are being taken.

p. 123

The school phobia study by Klein is cited in the Notes of Chapter 6.

p. 126

The improvement in musical performance ascribed to propranolol was reported in the 12 September, 1982, issue of the *New York Times*.

Index

abreaction 101
adrenaline 4, 6, 99
age of risk
  agoraphobia 63
  animal phobia 35
  depressive disorders 87
  heights, fear of 37
  simple phobias 40–1
  social phobias 54
agoraphobia 25, 28, 43, 55–63, 103, 119, 136–7
  in clinical depression 84
  treatment 119, 122–3
airline pilots, fear of heights in 37
alcohol 20, 21, 27, 28, 54, 61, 84, 120
Amiel 83
amnesia 100, 101–2
anger, reducing fear 22
animals, fear of 32, 33, 34–6
anorexia nervosa 49
antidepressants 73, 77, 87, 101
anxiety 1, 99, 127
  drugs for 6
anxiolytics 119
appetite, loss of 86
asthma 39
autogenic training 115
avoidance rituals 80

babies
  smile response in 8
  strangers, fear of in 8
Bacon, Sir Francis 16
barbiturates 20
behaviour therapy 106–18
belching 76
benzodiazepines 119, 144–5
beta-blockers 125–6
birth complications 67
blood, fear of 32, 52
blushing, fear of 52–3, 66
breathlessness 75
Bunyan, John 42–3
Burton, Robert 2

cancer, fear of 38
car driving, fear of 32, 53

castration anxiety 92
catharsis 101
causes of phobias 89–104
childhood, phobias in 7, 8, 9, 64–73, 116–18, 138; *see also* school phobia
cognitive therapy 113–14
claustrophobia 22, 32, 56
colitis 39
conditioning
  classical 15–18
  operant 18–20
Condon, Richard 85
constipation 76
contamination, fear of 35, 78, 79, 80
convictions, obsessional 78
counterphobias 23, 93
counting rituals 79–80
crowds, fear of 51–2

dark, fear of 7, 118
  woods 10–11
Darwin, Charles 2, 128
Davin, Dan 21–2
dead bodies, fear of 9
delusions 85–6
dentists, fear of 32, 33, 138
depersonalization 57, 75
depression 42, 51, 57, 61, 68, 72, 81
  drugs for 121–5
  see also depressive disorders
depressive disorders 83–8, 141
Descartes, René 99
diabetes 84
diarrhoea 76
dizziness 55–6
doctors, fear of 65
dogs, fear of 102, 116–17
drugs for phobias 118–26

eating
  fear of 44, 48–9
  reducing fear 23
electroconvulsive therapy 88
endorphins 129–30
envy 128
epilepsy 57
events, traumatic 67, 101–2

146

Index

evolutionary scale 13

fainting, fear of 52, 69
fear 1, 127, 130
   centres in brain 5, 129
   obsessional 79
flatus 76
flooding 111–12
flying, fear of 22, 25, 30, 105, 119–20, 133
France 102
Freud, Sigmund 29, 36, 45–6, 68, 89, 90–2, 105
Freudian fears 68
frigidity 50

gender differences 104
   and agoraphobia 62
   and animal phobias 35
   in childhood 65–6
   and obsessive-compulsive disorders 77
   and panic disorder 75
   and school phobia 69
   and simple phobias 40, 41
genetics 14; *see also* inherited fears
goats, fear of heights in 11
gossip 21
graded exposure 106–11
guilt 127–8

Haldane, J.B.S. 89
Hall, Stanley 10
hawk experiment 7, 130
heart disease
   and depression 84
   fear of 38, 102, 115
heart symptoms 75–6
Hebb, Donald 8
heights, fear of 11, 32, 36–7, 92, 109, 111
heredity 7–15, 35, 62, 97–100, 130
hierarchy, constructing 107
Hippocrates 28, 52
historical background 28–9
horses, fear of 90–2
Howard, Russell John 74
humour 114
Huxley, Aldous 24
hydrophobia 28, 132–3
hypnosis 105

hypochondriasis 39; *see also* illness, fear of

ideas, obsessional 78
illness, fear of 33, 38–9
images, obsessional 78
imipramine 121–4, 145
immobility 128–9
implosion therapy 112
impotence 50
imprinting 14
impulses, obsessional 79
incidence
   of agoraphobia 62
   of obsessive-compulsive disorder 77
   of phobia 26–7
inherited fears 7–15, 35
injections, fear of 32
innate fears 7–15, 97–100
'irritable colon' 76

Jacobsen, Edmund 108
James-Lange theory 6
James, William 6, 94, 125
Janet, Pierre 52
Jaspers, Karl 78
jealousy 128
Jong, Erica 30

Keller, Mark 17
Klein, Donald 122
Kraeplin, Emil 29

lactic acid 6, 99
lavatories, fear of public 45–8, 69
learned fears 15–20
Librium 5, 77, 119, 120
lightning, fear of 32, 65
Little Albert 94–6
Little Hans 90–2
Lorenz, Konrad 7, 12–13
LSD 10–11

mania 87
manic-depressive disease 87
MAO inhibitors 83, 121–4
Marks, Isaac 12, 37, 44, 58, 103, 127
Maudsley Hospital 31, 58
Mead, Margaret 46
Menninger, Karl 55, 60
meticulousness 81

Phobia: the facts

mitral valve prolapse 77
mother, overprotective 70, 71, 100
motion reducing fear 23

nausea 76
Nemiah, John 43, 52-3
nervous system 2-7
neuroleptics 119
noise, fear of 117-18
novelty, fear of 12-13

obsessive-compulsive disorders 77-83, 84

pain, tolerance of 5-6
palpitations 6
panic 1
  attacks 31, 40, 120, 122-3
  disorder 75-7
paradoxical intention 114-15
paraplegics 6
participant modelling 109-10, 112
Pavlov, I.P. 15, 17, 90
penis envy 90-2
personality of sufferers 99
phenelzine 121-4
physical contact, fear of 53
physical effects
  of fear 2-7, 39
  of school phobia 69
positive thinking 113-14
post-traumatic stress disorder 101
premature ejaculation 50
prevalence
  of agoraphobia 62
  of obsessive-compulsive disorder 77
  of phobia 26-7
propanolol 125-6
public speaking, fear of 44-5

quadraplegics 6

raven, experiment with 12
relaxation 108, 115
respiratory symptoms 76
rituals, obsessional 79-81
rumination, obsessional 79

schizophrenia 72, 84
school phobia 58, 69-73, 123, 139

Sergeant, William 121
sex
  and agoraphobia 57
  fear of 50
  reducing fear 22
sexes, differences between, see gender differences
shame 128
Sidney, Sir Philip 38-9
simple phobia 25, 30-53
Smith, Adam 10
snakes, fear of 9-10, 20, 34-5, 65, 107-8
social phobias 25, 44-54
sodium amytal 108
soteria 22-3
spiders, fear of 34-5, 65
stage fright 44-5, 125-6
staring, fear of 11-12, 66, 135; see also social phobias
state-dependent learning 123
storms, fear of 32, 64, 109
strangers, fear of 8-9, 65
suicide 77, 82, 84, 85, 87, 141
summation 15
swallowing, difficulty in 76
syphilis, fear of 28, 38, 134

talking as therapy 21
Thorazine 119
toilet training 45-8, 135-6
toilets, fear of public 45-8, 69
Torrey, E. Fuller 134
touch, fear of 53
tranquillizers 61, 77, 108, 119
trauma 67, 101-2
treatment 105-26
tricyclics 77, 121-4
Turner, Frederick J. 89
twins, identical 98-9

ungraded exposure 111-12
'urethral personality' 46
urinating, difficulty in
  in depressive disorders 86
  see also toilets, fear of public
USA 26, 102, 141

Valium 5, 77, 119, 120
venereal disease, fear of 28, 38, 134
vomiting, fear of 49-50

Index

watched, fear of being 11–12, 66; *see also* social phobias
Watson, John 89–90, 94
weight, loss of 86
Westphal, Karl 28, 55, 58, 60

Wolpe, Joseph 23, 31, 99–100, 105–18
woods, fear of dark 10–11
working, fear of being watched while 51
'writer's block' 51